ENGINEERING MECHANICS
DYNAMICS
R. C. HIBBELER

DYNAMICS STUDY PACK

FREE BODY DIAGRAM WORKBOOK
PETER SCHIAVONE

WORKING MODEL SIMULATION CD
GILBERT EMMERT

PROBLEMS WEBSITE
RUSSELL C. HIBBELER

PRENTICE HALL, Upper Saddle River, NJ 07458

Acquisitions Editor: Eric Svendsen
Supplement Editor: Kristen Blanco
Special Projects Manager: Barbara A. Murray
Production Editor: Barbara A. Till
Supplement Cover Manager: Paul Gourhan
Supplement Cover Designer: PM Workshop Inc.
Manufacturing Buyer: Lisa McDowell

© 2001 by Russell C. Hibbeler
Published by Prentice Hall
Upper Saddle River, NJ 07458

Printed in the United States of America

10 9 8 7 6 5 4 3 2 1

ISBN   0-13-090757-X

Prentice-Hall International (UK) Limited, London
Prentice-Hall of Australia Pty. Limited, Sydney
Prentice-Hall Canada, Inc., Toronto
Prentice-Hall Hispanoamericana, S.A., Mexico
Prentice-Hall of India Private Limited, New Delhi
Pearson Education Asia Pte. Ltd., Singapore
Prentice-Hall of Japan, Inc., Tokyo
Editora Prentice-Hall do Brazil, Ltda., Rio de Janeiro

Foreword

The Dynamics Study Pack was designed to help students improve their study skills. It consists of three study components — a free body diagram workbook, a Visualization CD based on Working Model Software, and an access code to a website with over 1000 sample Statics and Dynamics problems and solutions.

- **Free Body Diagram Workbook** — Prepared by Peter Schiavone of the University of Alberta. This workbook begins with a tutorial on free body diagrams and then includes 50 practice problems of progressing difficulty with complete solutions. Further "strategies and tips" help students understand how to use the diagrams in solving the accompanying problems.

- **Working Model CD** — Prepared with the help of Gil Emmert of the University of Wisconsin, Madison. This CD contains 90 pre-set simulations of Dynamics examples in the text that include questions for further exploration. Simulations are powered by the Working Model Engine and were created with actual artwork from the text to enhance their correlation with the text. Directions for CD installation are on the CD's README file. You need to have the CD in your drive when using the simulations. Please also note the licensing terms for using the CD.

- **Problems Website** — Located at http://www.prenhall.com/hibbeler. This Website contains 1000 sample Statics and Dynamics problems for students to study. Problems are keyed to each chapter of the text and contain complete solutions. All problems are supplemental and do not appear in the Eighth or Ninth Edition. Student passwords are printed on the inside cover of the Free Body Diagram Workbook. To access this site, students should go to http://www.prenhall.com/hibbeler, choose the link for the Problems Website, and follow the on-line directions to register. This site also contains an unprotected section with multiple choice check up questions.

Preface

*A thorough understanding of how to draw and use a free-body diagram
is absolutely essential when solving problems in mechanics.*

This workbook consists mainly of a collection of problems intended to give the student practice in drawing and using free-body diagrams when solving problems in dynamics.

All the problems are presented as tutorial problems with the solution only partially complete. The student is then expected to complete the solution by 'filling in the blanks' in the spaces provided. This gives the student the opportunity to build free-body diagrams in stages and extract the relevant information from them when formulating equations of motion. Earlier problems provide students with partially drawn free-body diagrams and lots of hints to complete the solution. Later problems are more advanced and are designed to challenge the student more. The complete solution to each problem can be found at the back of the page. The problems are chosen from two-dimensional theories of particle and rigid body dynamics. Once the ideas and concepts developed in these problems have been understood and practiced, the student will find that they can be extended in a relatively straightforward manner to accommodate the corresponding three-dimensional theories.

The book begins with a brief primer on free-body diagrams: where they fit into the general procedure of solving problems in dynamics and why they are so important. Next follows a few examples to illustrate ideas and then the workbook problems.

For best results, the student should read the primer and then, beginning with the simpler problems, try to complete and understand the solution to each of the subsequent problems. The student should avoid the temptation to immediately look at the completed solution over the page. This solution should be accessed only as a last resort (after the student has struggled to the point of giving up), or to check the student's own solution after the fact. The idea behind this is very simple: we learn most when we *do* the thing we are trying to learn - reading through someone else's solution is not the same as actually working through the problem. In the former, the student gains information, in the latter the student gains knowledge. For example, how many people learn to swim or drive a car by reading an instruction manual?

Consequently, since this book is based on *doing*, the student who persistently solves the problems in this book will ultimately gain a thorough, usable knowledge of how to draw and use free-body diagrams.

P. Schiavone

Contents

1

Basic Concepts in Dynamics

Engineering mechanics is divided into two areas: statics and dynamics. *Statics* deals with the equilibrium of bodies, that is, those that are either at rest (if originally at rest) or move with constant velocity (if originally in motion). *Dynamics* is concerned with the accelerated motion of bodies. The study of dynamics is itself divided into two parts: *kinematics*, which treats only the geometric aspects of motion and *kinetics* which is concerned with the analysis of forces causing the motion. Free-body diagrams play a significant role in solving problems in kinetics.

In mechanics, real bodies (e.g. planets, cars, planes, tables, crates, etc) are represented or *modeled* using certain idealizations which simplify application of the relevant theory. In this book we refer to only two such models:

- **Particle**. A *particle* has a mass but a size/shape that can be neglected. For example, the size of an aircraft is insignificant when compared to the size of the earth and therefore the aircraft can be modeled as a particle when studying its three-dimensional motion in space.
- **Rigid Body**. A *rigid body* represents the next level of sophistication after the particle. That is, a rigid body is a collection of particles which has a size/shape but this size/shape cannot change. In other words, when a body is modeled as a rigid body, we assume that any deformations (changes in shape) are relatively small and can be neglected. For example, the actual deformations occurring in most structures and machines are relatively small so that the rigid body assumption is suitable in these cases.

1.1 Equations of Motion

Equation of Motion for a Particle
When a system of forces acts on a particle, the equation of motion may be written in the form

$$\sum \mathbf{F} = m\mathbf{a} \tag{1.1}$$

where $\sum \mathbf{F}$ is the vector sum of all the external forces acting on the particle and m and $\mathbf{a}$ are, respectively, the mass and acceleration of the particle.

Successful application of the equation of motion (1.1) requires a complete specification of all the known and unknown external forces ($\sum \mathbf{F}$) that act on the object. The best way to account for these is to draw the object's *free-body diagram*: a sketch of the object *freed* from its surroundings showing *all* the (external) forces that act on it. In dynamics problems, since the resultant of these external forces produces the vector $m\mathbf{a}$, in addition to the free-body diagram, a *kinetic diagram* is often used to represent graphically the magnitude and direction of the vector $m\mathbf{a}$. In other words, the equation (1.1) can be represented graphically as:

Free-body Diagram = Kinetic Diagram

Of course, whenever the equation of motion (1.1) is applied, it is required that measurements of the acceleration be made from a *Newtonian* or inertial frame of reference. *Such a coordinate system does not rotate and is either fixed or translates in a given direction with a constant velocity (zero acceleration).* This definition ensures that the particle's acceleration measured by observers in two different inertial frames of reference will always be the *same*.

Equation of Motion for a System of Particles

The equation of motion (1.1) can be extended to include a *system of particles* isolated within an enclosed region in space:

$$\sum \mathbf{F} = m\mathbf{a}_G \tag{1.2}$$

This equation states that the sum of external forces ($\sum \mathbf{F}$) acting on the system of particles is equal to the total mass m of the particles multiplied by the acceleration $\mathbf{a}_G$ of its mass center G. Since, in reality, all particles must have a finite size to possess mass, equation (1.2) justifies application of the equation of motion to a *body* that is represented as a single particle.

Equations of Motion for a Rigid Body

Since rigid bodies, by definition, have a definite size/shape, their motion is governed by *both* translational and rotational quantities. The translational equation of motion for (the mass center of) a rigid body is basically equation (1.2). That is,

$$\sum \mathbf{F} = m\mathbf{a}_G \tag{1.2}$$

In this case, the equation (1.2) states that the sum of all the external forces acting on the body is equal to the body's mass multiplied by the acceleration of its mass center G.

The rotational equation of motion for a rigid body is given by

$$\sum \mathbf{M}_G = I_G \alpha \tag{1.3}$$

which states that the sum of the applied couple moments and moments of all the external forces computed about a body's mass center $G(\sum \mathbf{M}_G)$ is equal to the product of the moment of inertia of the body about an axis passing through $G(I_G)$ and the body's angular acceleration α.

Alternatively, equation (1.3) can be re-written in more general form as:

$$\sum \mathbf{M}_P = \sum (M_k)_P \tag{1.4}$$

Here, $\sum \mathbf{M}_P$ represents the sum of the applied couple moments and the external moments taken about a general point P and $\sum (M_k)_P$ represents the sum of the kinetic moments about P, in other words, the sum of $I_G \alpha$ and the moments generated by the components of the vectors $m\mathbf{a}_G$ about the point P.

When applying the equations of motion (1.2)–(1.4), one should always draw a *free-body diagram* in order to account for the terms involved in ($\sum \mathbf{F}$), ($\sum \mathbf{M}_G$) or ($\sum \mathbf{M}_P$). The *kinetic diagram* is also useful in that it accounts graphically for the acceleration components $m\mathbf{a}_G$ and the term $I_G \alpha$ and it is especially convenient when used to determine the components of $m\mathbf{a}_G$ and the moment terms in $\sum (M_k)_P$.

2

Free-Body Diagrams: the Basics

2.1 Free-Body Diagram: Particle

The equation of motion (1.1) is used to analyze the motion of *an object* (modeled as a particle) when subjected to an unbalanced force system. The first step in this analysis is to draw the **free-body diagram** of the object to identify the external forces ($\sum \mathbf{F}$) acting on it. The object's free-body diagram is simply a sketch of the object *freed* from its surroundings showing *all* the (external) forces that *act* on it. The diagram focuses your attention on the object of interest and helps you identify *all* the external forces ($\sum \mathbf{F}$) acting. Once the free-body diagram is drawn, it may be helpful to draw the corresponding *kinetic diagram*. This diagram accounts graphically for the effect of the acceleration components ($m\mathbf{a}$) on the object. Taken together, these diagrams provide (in graphical form) all the information that is needed to write down the equation of motion (1.1).

EXAMPLE 2.1

The 50-kg crate shown in Figure 1, rests on a horizontal plane for which the coefficient of friction is $\mu_k = 0.3$. The crate is subjected to a towing force of magnitude 400N and moves to the right without tipping over. Draw the free-body and kinetic diagrams of the crate.

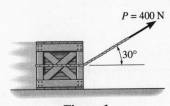

$P = 400$ N

$30°$

Figure 1

Solution

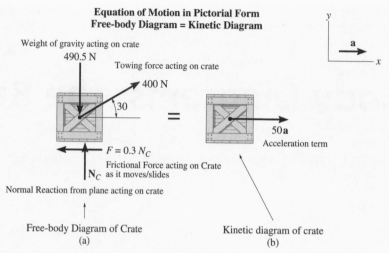

Equation of Motion in Pictorial Form
Free-body Diagram = Kinetic Diagram

Weight of gravity acting on crate
490.5 N

Towing force acting on crate
400 N

30

=

50**a**
Acceleration term

$F = 0.3 N_C$
Frictional Force acting on Crate
N_C as it moves/slides

Normal Reaction from plane acting on crate

Free-body Diagram of Crate
(a)

Kinetic diagram of crate
(b)

Figure 2

The free-body diagram of the crate is shown in Figure 2(a). Notice that once the crate is *separated* or *freed* from the system (= crate + plane), forces which were previously internal to the system become external to the crate. For example, in Figure 2 (a), such a force is the force of friction *acting on the crate*. The kinetic diagram is shown in Figure 2(b). In this case, the diagram shows the effect of the acceleration term $m\mathbf{a}$ on the crate. Taken together, the two diagrams give a pictorial form of the equation of motion (1.1) (or (1.2)).   ◀

Next, we present a formal procedure for drawing free-body diagrams for a particle or system of particles.

2.1.1 Procedure for Drawing a Free-Body Diagram: Particle

1. *Select* the inertial coordinate system. Most often, rectangular or x, y-coordinates are chosen to analyze problems for which the particle has *rectilinear motion*. If this occurs, one of the axes should extend in the direction of motion.

2. *Identify the object you wish to isolate* from the system. This choice is often dictated by the particular forces of interest.

3. *Draw the outlined shape of the isolated object*. Imagine the object to be isolated or cut free from the system of which it is a part.

4. *Show all external forces acting on the isolated object*. Indicate on this sketch *all* the external forces that act on the object. These forces can be *active forces*, which tend to set the object in motion, or they can be *reactive forces* which are the result of the constraints or supports that prevent motion. This stage is crucial: it may help to trace around the object's boundary, carefully noting each external force acting on it. Don't forget to include the weight of the object (unless it is being intentionally neglected).

5. *Identify and label each external force acting on the (isolated) object*. The forces that are known should be labeled with their known magnitudes and directions. Use letters to represent the magnitudes and arrows to represent the directions of forces that are unknown.

6. *The direction of a force having an unknown magnitude can be assumed*.

7. *The direction and sense* of the particle's acceleration **a** should also be established. If the sense of its components is unknown, assume they are in the same direction as the positive inertial coordinate axes. The acceleration may be sketched on the x, y-coordinate system or it may be represented as the $m\mathbf{a}$ vector on the *kinetic diagram*.

2.1.2 Using the Free-Body Diagram: Equations of Motion

The equations of motion (1.1) or (1.2) are used to solve problems which require a relationship between the forces acting on a particle and the accelerated motion they cause. Whenever (1.1) or (1.2) is applied, the unknown force and acceleration components should be identified and an equivalent number of equations should be written. If further equations are required for the solution, kinematics may be considered.

The *free-body diagram* is used to identify the unknown force and the *kinetic diagram* the unknown acceleration components acting on the particle. The subsequent procedure for solving problems once the free-body (and, if necessary, the kinetic) diagram for the particle is established, is therefore as follows:

---

1. If the forces can be resolved directly from the free-body diagram, apply the equations of motion in their scalar component form. For example:

$$\sum F_x = ma_x \quad \text{and} \quad \sum F_y = ma_y \tag{2.1}$$

2. Components are positive if they are directed along a positive axis and negative if they are directed along a negative axis.

3. If the particle contacts a rough surface, it may be necessary to use the frictional equation, which relates the coefficient of kinetic friction to the magnitudes of the frictional and normal forces acting at the surfaces of contact. Remember that the frictional force always acts on the free-body diagram such that it opposes the motion of the particle *relative to the surface it contacts*.

4. If the solution yields a negative result, this indicates the sense of the force is the reverse of that shown/assumed on the free-body diagram.

EXAMPLE 2.2

In Example 2.1, the diagrams established in Figure 2 give us a 'pictorial representation' of all the information we need to apply the equations of motion (2.1) to find the unknown force N_C and the acceleration **a**. In fact, taking the positive x-direction to be horizontal ($\rightarrow +$) and the positive y-direction to be vertical ($\uparrow +$), the equations of motion (2.1) when applied to the crate (regarded as a particle — since its shape is not important in the motion under consideration) are:

For the Crate: $\rightarrow + \sum F_x = ma_x$: $400 \cos 30° - F = 50a_x$
$\uparrow + \sum F_y = ma_y$: $N_C - 490.5 + 400 \sin 30° = 0$

Two equations, 3 unknowns: use the frictional equation to relate F to N_C and obtain a third equation:

Frictional Equation (block is sliding): $F = 0.3N_C$

Solving these three equations yields

$$N_C = 290.5\text{N}, \quad a_x = 5.19 m/s^2 \qquad \textbf{Ans.}$$

The directions of each of the vectors N_C and **a** is shown in the free-body diagram above (Figure 2).   ◄

2.2 Free-Body Diagram: Rigid Body

The equations of motion (1.2) and (1.3) (or (1.4)) are used to determine unknown forces, moments and acceleration components acting on an object (modeled as a rigid body) subjected to an unbalanced system of forces and moments. The first step in doing this is again to draw the *free-body diagram* of the object to identify *all of* the external forces and moments acting on it. The procedure for drawing a free-body diagram in this case is much the same as that for a particle with the main difference being that now, because the object has 'size/shape,' it can support also external couple moments and moments of external forces.

2.2.1 Procedure for Drawing a Free-Body Diagram: Rigid Body

1. *Select* the inertial x, y or n, t-coordinate system. This will depend on whether the body is in rectilinear or curvilinear motion.
2. Imagine the body to be isolated or 'cut free' from its constraints and connections and sketch its outlined shape.
3. Identify all the external forces and couple moments that act on the body. Those generally encountered are:
   (a) Applied loadings
   (b) Reactions occurring at the supports or at points of contact with other bodies.
   (c) The weight of the body (applied at the body's center of gravity G)
   (d) Frictional forces
4. The forces and couple moments that are known should be labeled with their proper magnitudes and directions. Letters are used to represent the magnitudes and direction angles of forces and couple moments that are *unknown*. Indicate the dimensions of the body necessary for computing the moments of external forces. In particular, if a force or couple moment has a known line of action but unknown magnitude, the arrowhead which defines the sense of the vector can be assumed. The correctness of the assumed sense will become apparent after solving the equations of motion for the unknown magnitude. By definition, the magnitude of a vector is *always positive*, so that if the solution yields a *negative* scalar, the minus *sign* indicates that the vector's sense is *opposite* to that which was originally assumed.
5. *The direction and sense* of the acceleration of the body's mass center $\mathbf{a}_G$ should also be established. If the sense of its components is unknown, assume they are in the same direction as the positive inertial coordinate axes. The acceleration may be sketched on the x, y-coordinate system or it may be represented as the $m\mathbf{a}_G$ vector on the *kinetic diagram*. This will also be helpful for 'visualizing' the terms needed in the moment sum $\sum (M_k)_P$ since the kinetic diagram accounts graphically for the components $m(a_G)_x$, $m(a_G)_x$ or $m(a_G)_t$, $m(a_G)_n$.

2.2.2 Important Points

- Internal forces are never shown on the free-body diagram since they occur in equal but opposite collinear pairs and therefore cancel each other out.
- The weight of a body is an external force and its effect is shown as a single resultant force acting through the body's center of gravity G.
- *Couple moments* can be placed anywhere on the free-body diagram since they are *free vectors*. Forces can act at any point along their lines of action since they are *sliding vectors*.

EXAMPLE 2.3

Draw the free-body and kinetic diagrams for the 50-kg crate. A force **P** of magnitude $600N$ is applied to the crate as shown. Take the coefficient of kinetic friction to be $\mu_k = 0.2$.

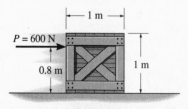

Figure 3

Solution

Here, since the force **P** can cause the crate to either slide or to tip over, we model the crate as a rigid body. This model allows us to account for the effects of moments arising from **P** and any other external forces. We begin by assuming that the crate slides so that the frictional equation yields $F = \mu_k N_C = 0.2N_C$. Also, the normal force N_C acts at O, a distance x (where $0 < x \leq 0.5m$) from the crate's center line. Note that the line of action of N_C does not necessarily pass through the mass center $G(x = 0)$, since N_C must counteract the tendency for tipping caused by **P**.

(Note that had we assumed that the crate tips then the normal force N_C would have been assumed to act at the corner point A and the frictional equation would take the form $F \leq 0.2N_C$).   ◄

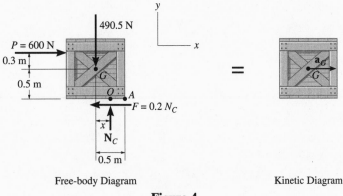

Free-body Diagram   Kinetic Diagram

Figure 4

2.2.3 Using the Free-Body Diagram: Equations of Motion

The procedure for solving kinetic problems for a rigid body once the free-body diagram is established, is as follows:

- Apply the three equations of motion (1.2)–(1.3). To simplify the analysis, the moment equation (1.3) may be replaced by the more general equation (1.4) where the point P is usually located at the intersection of the lines of action of as many unknown forces as possible.
- If the body contacts a rough surface, it may be necessary to use the frictional equation, which relates the coefficient of kinetic friction to the magnitudes of the frictional and normal forces acting at the surfaces of contact. Remember that the frictional force always acts on the free-body diagram such that it *opposes the motion of the body relative to the surface it contacts*.
- Use kinematics if the velocity and position of the body are to be determined.

EXAMPLE 2.4

Find the acceleration of the crate in Example 2.3.

Solution

Using the free-body diagram in Figure 4, the equations of motion are:

$$\longrightarrow + \sum F_x = m(a_G)_x: \ 600N - 0.2N_C = (50kg)(a_G)_x$$
$$\uparrow + \sum F_y = m(a_G)_y: \ N_C - 490.5N = 0$$
$$+ \circlearrowleft \sum M_G = I_G\alpha: \ -600N(0.3m) + N_C(x) - 0.2N_C(0.5m) = 0$$

Solving, we obtain

$$\mathbf{N}_C = 490N \uparrow, \quad x = 0.467m, \quad \mathbf{a}_G = 10m/s^2 \longrightarrow \qquad \qquad \textbf{Ans.}$$

Since $x = 0.467m < 0.5m$, indeed the crate slides as originally assumed (otherwise the problem would have to be reworked with the assumption that tipping occurred).   ◀

3

Problems

3.1 Free-Body Diagrams in Particle Kinetics

Problem 3.1

The sled with load shown has a weight of 50 lb and is acted upon by a force having a variable magnitude $P = 20t$, where P is in pounds and t is in seconds. The coefficient of kinetic friction between the sled and the plane is $\mu_k = 0.3$. Draw the free-body and kinetic diagrams for the sled.

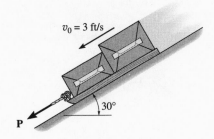

Solution

1. The size/shape of the sled does not affect the (rectilinear) motion under consideration. Consequently, we assume that the sled has *negligible size* so that it can be modelled as a particle.
2. Imagine the sled to be separated or detached from the system (sled + plane).
3. The (detached) sled is subjected to four *external* forces. They are caused by:

   i.   **ii.**

   iii.   **iv.**

4. Draw the free-body diagram of the (detached) sled showing all these forces labeled with their magnitudes and directions. Include any other information e.g. angles, lengths etc. which may help when formulating the equations of motion.
5. The acceleration of the sled is down the slope. Show this on a kinetic diagram or on the inertial coordinate system chosen in the free-body diagram.

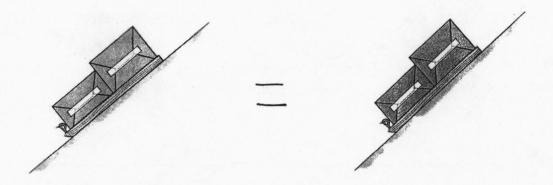

Problem 3.1

The sled with load shown has a weight of 50 lb and is acted upon by a force having a variable magnitude $P = 20t$, where P is in pounds and t is in seconds. The coefficient of kinetic friction between the sled and the plane is $\mu_k = 0.3$. Draw the free-body and kinetic diagrams for the sled.

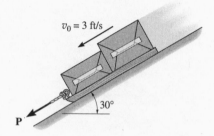

Solution

1. The size/shape of the sled does not affect the (rectilinear) motion under consideration. Consequently, we assume that the sled has *negligible size* so that it can be modelled as a particle.

2. Imagine the sled to be separated or detached from the system (sled + plane).

3. The (detached) sled is subjected to four *external* forces. They are caused by:

     i. Force P   **ii. Sled's weight**

    iii. Frictional force at the surface (sliding)   **iv. Force of surface acting on sled**

4. Draw the free-body diagram of the (detached) sled showing all these forces labeled with their magnitudes and directions. Include any other information e.g. angles, lengths etc. which may help when formulating the equations of motion.

5. The acceleration of the sled is down the slope. Show this on a kinetic diagram or on the inertial coordinate system chosen in the free-body diagram.

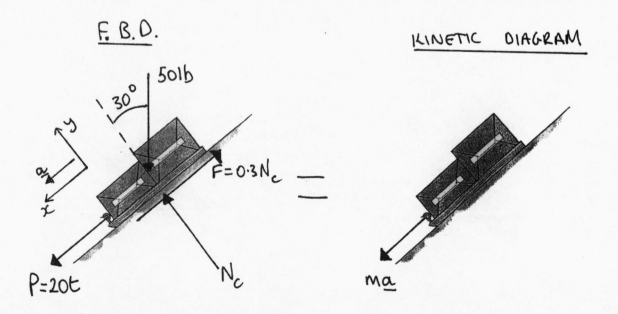

Problem 3.2

The man weighs 180 lb and supports the barbells which have a weight of 100 lb. He lifts the barbells 2 ft in the air in 1.5 secs. Draw the free-body and kinetic diagrams for the man holding the barbells.

Solution

1. The size/shape of the man with barbells does not affect the (rectilinear) motion under consideration. Consequently, we assume that the man with barbells has *negligible size* so that together they can be modelled as a particle.

2. Imagine the man with barbells to be separated or detached from the system (man with barbells + ground).

3. The (detached) man with barbells is subjected to three *external* forces. They are caused by:

   i.   **ii.**

   iii.

4. Draw the free-body diagram of the (detached) man with barbells showing all these forces labeled with their magnitudes and directions. Include any other information e.g. angles, lengths etc. which may help when formulating the equations of motion.

5. The acceleration of the barbells is upward. Show this on a kinetic diagram or on the inertial coordinate system chosen in the free-body diagram.

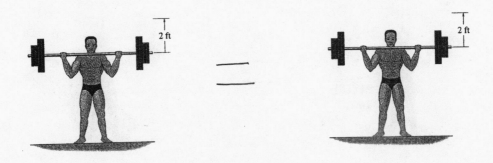

Problem 3.2

The man weighs 180 lb and supports the barbells which have a weight of 100 lb. He lifts the barbells 2 ft in the air in 1.5 secs. Draw the free-body and kinetic diagrams for the man holding the barbells.

Solution

1. The size/shape of the man with barbells does not affect the (rectilinear) motion under consideration. Consequently, we assume that the man with barbells has *negligible size* so that together they can be modelled as a particle.

2. Imagine the man with barbells to be separated or detached from the system (man with barbells + ground).

3. The (detached) man with barbells is subjected to three *external* forces. They are caused by:

    i. Man's weight   **ii. Weight of barbells**

    iii. (Single) reaction of floor to man (recall man is modelled as a *particle*)

4. Draw the free-body diagram of the (detached) man with barbells showing all these forces labeled with their magnitudes and directions. Include any other information e.g. angles, lengths etc. which may help when formulating the equations of motion.

5. The acceleration of the barbells is upward. Show this on a kinetic diagram or on the inertial coordinate system chosen in the free-body diagram.

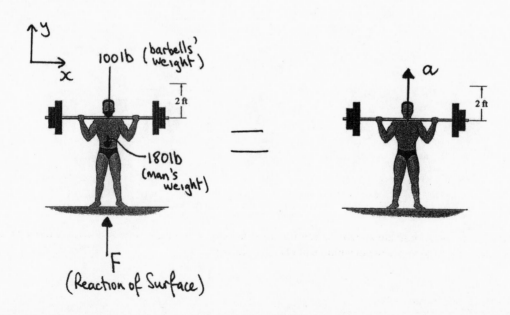

Problem 3.3

The water-park ride consists of an 800-lb sled which slides from rest down the incline and then into the pool. If the frictional resistance on the incline is $F_r = 30$ lb and, in the pool for a short distance $F_r = 80$ lb, draw the free-body and kinetic diagrams for the sled (a) on the incline (b) in the pool. Use these diagrams to determine how fast the sled is travelling when $s = 5$ ft.

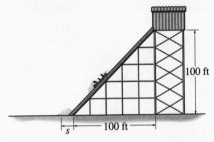

100 ft

100 ft

s

Solution

1. The size/shape of the sled does not affect the (rectilinear) motion under consideration. Consequently, we assume that the sled has *negligible size* so that it can be modelled as a particle.
2. Imagine the sled to be separated or detached from the system (sled + inclined plane or sled + pool).
3. In each case, the (detached) sled is subjected to three *external* forces. They are caused by:

   i.   ii.

   iii.

4. Draw the free-body and kinetic diagrams of the (detached) sled showing all these forces labeled with their magnitudes and directions for the sled (a) on incline (b) in pool. Include any other information e.g. angles, lengths etc. which may help when formulating the equations of motion.

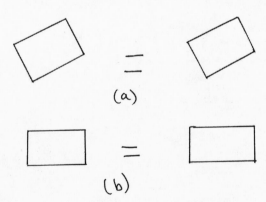

5. In each case, using the coordinate system chosen for the free-body diagrams, write down the equations of motion in the direction of motion and solve for the magnitude of the acceleration:

   Slope: $+\swarrow \sum F_x = ma_x$:

   Pool: $\longleftarrow +\sum F_x = ma_x$:

6. Use kinematics to find the speed of the sled at $s = 5$ ft.

Problem 3.3

The water-park ride consists of an 800-lb sled which slides from rest down the incline and then into the pool. If the frictional resistance on the incline is $F_r = 30$ lb and, in the pool for a short distance $F_r = 80$ lb, draw the free-body and kinetic diagrams for the sled (a) on the incline (b) in the pool. Use these diagrams to determine how fast the sled is travelling when $s = 5$ ft.

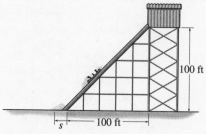

Solution

1. The size/shape of the sled does not affect the (rectilinear) motion under consideration. Consequently, we assume that the sled has *negligible size* so that it can be modelled as a particle.
2. Imagine the sled to be separated or detached from the system (sled + inclined plane or sled + pool).
3. In each case, the (detached) sled is subjected to three *external* forces. They are caused by:

   i. Action of surface on sled   **ii. Sled's weight**

   iii. Frictional force at the surface (sliding)

4. Draw the free-body and kinetic diagrams of the (detached) sled showing all these forces labeled with their magnitudes and directions for the sled (a) on incline (b) in pool. Include any other information e.g. angles, lengths etc. which may help when formulating the equations of motion.

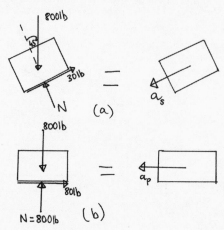

5. In each case, using the coordinate system chosen for the free-body diagrams, write down the equations of motion in the direction of motion and solve for the magnitude of the acceleration:

   Slope: $+ \swarrow \sum F_x = ma_x$: $800 \sin 45° - 30 = \dfrac{800 a_s}{32.2} \implies a_s = 21.561$ ft/s^2

   Pool: $\longleftarrow + \sum F_x = ma_x$: $-80 = \dfrac{800 a_p}{32.2} \implies a_p = -3.22$ ft/s^2

6. Use kinematics to find the speed of the sled at $s = 5$ ft:

   Upon entering the water, the sled has speed v_1 such that $v_1^2 = v_0^2 + 2a_s(s - s_0) = 0 + 2(21.561)\sqrt{20000} = 78.093$ ft/s.
   At $s = 5$ ft, (i.e. 5 ft into the pool)

   $$v_2^2 = v_1^2 + 2a_p(s_2 - s_1) = (78.093)^2 + 2(-3.22)(5 - 0) = 6068.4 \text{ ft/s}$$
   $$v_2 = 77.9 \text{ ft/s}$$

   Ans.

Problem 3.4

Each of the two blocks has a mass m. The coefficient of kinetic friction at all surfaces of contact is μ. A horizontal force **P** is applied to the bottom block. Draw free-body diagrams for each of the top and bottom blocks.

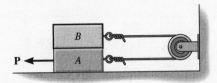

Solution

1. The size/shape of the blocks does not affect the motion under consideration. Consequently, we assume that the blocks have *negligible size* so that they can be modelled as particles.

2. Imagine each block to be separated or detached from the system (two blocks + plane).

3. The (detached) upper block is subjected to four *external* forces. They are caused by:

   i.   **ii.**

   iii.   **iv.**

   The (detached) lower block is subjected to six external forces. They are caused by:

   i.   **ii.**

   iii.   **iv.**

   v.   **vi.**

4. Draw the free-body diagrams of each (detached) block showing all these forces labeled with their magnitudes and directions. Include any other information e.g. angles, lengths etc. which may help when formulating the equations of motion.

5. What is the direction of the acceleration of each block? Show this on a kinetic diagram or on the inertial coordinate system chosen in the free-body diagram.

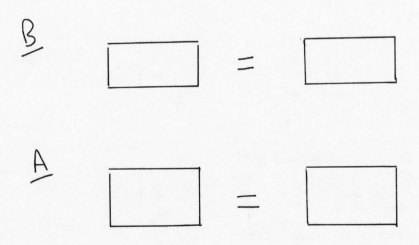

Problem 3.4

Each of the two blocks has a mass m. The coefficient of kinetic friction at all surfaces of contact is μ. A horizontal force **P** is applied to the bottom block. Draw free-body diagrams for each of the top and bottom blocks.

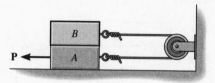

Solution

1. The size/shape of the blocks does not affect the motion under consideration. Consequently, we assume that the blocks have *negligible size* so that they can be modelled as particles.
2. Imagine each block to be separated or detached from the system (two blocks + plane).
3. The (detached) upper block is subjected to four *external* forces. They are caused by:

   i. It's weight
   ii. Cable Tension T
   iii. Friction between blocks
   iv. Reaction from lower block

   The (detached) lower block is subjected to six external forces. They are caused by:

   i. It's weight
   ii. Force P
   iii. Friction at supporting surface
   iv. Friction with upper block
   v. Reaction from surface
   vi. Cable Tension T

4. Draw the free-body diagrams of each (detached) block showing all these forces labeled with their magnitudes and directions. Include any other information e.g. angles, lengths etc. which may help when formulating the equations of motion.
5. What is the direction of the acceleration of each block? Show this on a kinetic diagram or on the inertial coordinate system chosen in the free-body diagram.

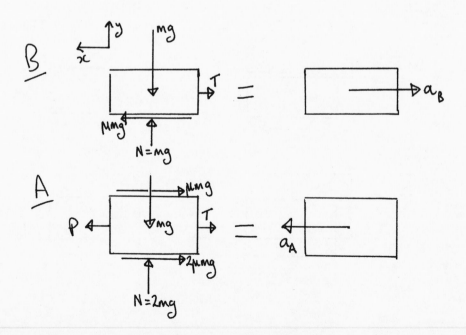

Problem 3.5

The driver attempts to tow the crate which has a weight of 500 lb and which is originally at rest. The coefficient of static friction between the crate and the ground is $\mu_s = 0.4$ and the coefficient of kinetic friction is $\mu_k = 0.3$. Draw free-body and kinetic diagrams for the crate just before and just after it begins to slide.

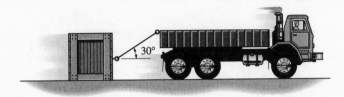

Solution

1. The size/shape of the crate does not affect the (rectilinear) motion under consideration. Consequently, we assume that the crate has *negligible size* so that it can be modelled as a particle.

2. Imagine the crate to be separated or detached from the system (crate + truck + ground).

3. In each case, the (detached) crate is subjected to four *external* forces. They are caused by:

    i.   **ii.**

    iii.   **iv.**

4. Draw the free-body diagram of the (detached) crate in each case, showing all these forces labeled with their magnitudes and directions. Include any other information e.g. angles, lengths etc. which may help when formulating the equations of motion.

5. Draw the corresponding kinetic diagram.

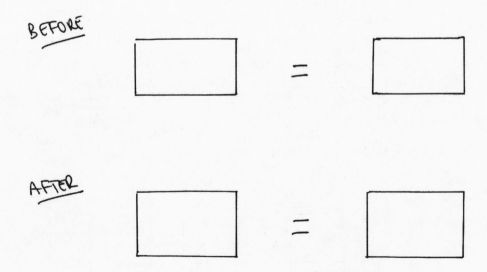

Problem 3.5

The driver attempts to tow the crate which has a weight of 500 lb and which is originally at rest. The coefficient of static friction between the crate and the ground is $\mu_s = 0.4$ and the coefficient of kinetic friction is $\mu_k = 0.3$. Draw free-body and kinetic diagrams for the crate just before and just after it begins to slide.

Solution

1. The size/shape of the crate does not affect the (rectilinear) motion under consideration. Consequently, we assume that the crate has *negligible size* so that it can be modelled as a particle.

2. Imagine the crate to be separated or detached from the system (crate + truck + ground).

3. In each case, the (detached) crate is subjected to four *external* forces. They are caused by:

   i. It's weight
   iii. Friction
   ii. Tension in rope
   iv. Reaction from surface

4. Draw the free-body diagram of the (detached) crate in each case, showing all these forces labeled with their magnitudes and directions. Include any other information e.g. angles, lengths etc. which may help when formulating the equations of motion.

5. Draw the corresponding kinetic diagram.

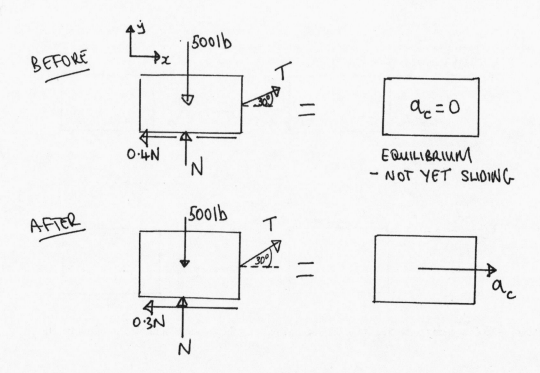

Problem 3.6

The block at B has a mass of 10 kg. Assume the surface at B is smooth. Neglect the mass of the pulleys and cords. Draw free-body and kinetic diagrams for the block at B and use them to formulate an equation of motion which gives a relationship between the acceleration of the block and the tension in the rope attached to B.

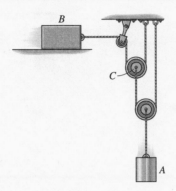

Solution

1. The size/shape of the block does not affect the (rectilinear) motion under consideration. Consequently, we assume that the block has *negligible size* so that it can be modelled as a particle.

2. Imagine the block to be separated or detached from the system.

3. The (detached) block is subjected to three *external* forces. They are caused by:

   i.   **ii.**

   iii.

4. Draw the free-body diagram of the (detached) block showing all these forces labeled with their magnitudes and directions. Include any other information e.g. angles, lengths etc. which may help when formulating the equations of motion.

5. Draw the corresponding kinetic diagram.

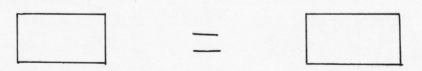

6. Using the $xy-axes$ system on the free-body diagram, write down the equation of motion in the x-direction:

$$\underset{\rightarrow}{+}\sum F_x = ma_x:$$

7. Solve for the acceleration of the block:

Problem 3.6

The block at B has a mass of 10 kg. Assume the surface at B is smooth. Neglect the mass of the pulleys and cords. Draw free-body and kinetic diagrams for the block at B and use them to formulate an equation of motion which gives a relationship between the acceleration of the block and the tension in the rope attached to B.

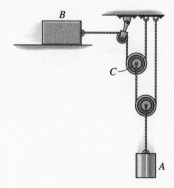

Solution

1. The size/shape of the block does not affect the (rectilinear) motion under consideration. Consequently, we assume that the block has *negligible size* so that it can be modelled as a particle.
2. Imagine the block to be separated or detached from the system.
3. The (detached) block is subjected to three *external* forces. They are caused by:

   i. It's weight   **ii. Tension in rope**

   iii. Reaction from surface

4. Draw the free-body diagram of the (detached) block showing all these forces labeled with their magnitudes and directions. Include any other information e.g. angles, lengths etc. which may help when formulating the equations of motion.
5. Draw the corresponding kinetic diagram.

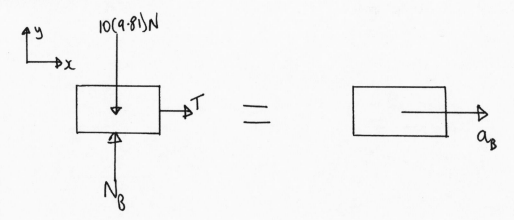

6. Using the xy-axes system on the free-body diagram, write down the equation of motion in the x-direction:

$$\underset{\rightarrow}{+} \sum F_x = ma_x: \quad T = 10a_B$$

7. Solve for the acceleration of the block:

$$a_B = \frac{T}{10} \text{ m/s}^2 \rightarrow \qquad\qquad\qquad \textbf{Ans.}$$

Problem 3.7

The 400-kg mine car is hoisted up the incline using the cable and motor M. For a short time, the force in the cable has magnitude $F = (3200t^2)N$ where t is in seconds. Draw a free-body diagram for the car and use it to determine the acceleration of the car after 2 seconds.

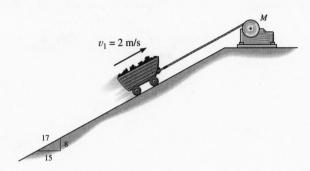

Solution

1. The size/shape of the car does not affect the (rectilinear) motion under consideration. Consequently, we assume that the car has *negligible size* so that it can be modelled as a particle.
2. Imagine the car to be separated or detached from the system.
3. The (detached) car is subjected to three *external* forces. They are caused by:

   i.   **ii.**

   iii.

4. Draw the free-body diagram of the (detached) car showing all these forces labeled with their magnitudes and directions. Include any other information e.g. angles, lengths etc. which may help when formulating the equations of motion.
5. Draw the corresponding kinetic diagram.

6. Using the $xy - axes$ system on the free-body diagram, write down the equation of motion in the x-direction:

$$\nearrow + \sum F_x = ma_x:$$

7. Solve for the acceleration of the car:

Problem 3.7

The 400-kg mine car is hoisted up the incline using the cable and motor M. For a short time, the force in the cable has magnitude $F = (3200t^2)\,N$ where t is in seconds. Draw a free-body diagram for the car and use it to determine the acceleration of the car after 2 seconds.

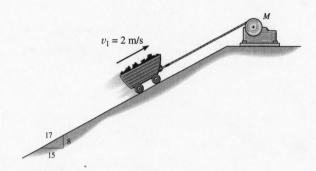

Solution

1. The size/shape of the car does not affect the (rectilinear) motion under consideration. Consequently, we assume that the car has *negligible size* so that it can be modelled as a particle.
2. Imagine the car to be separated or detached from the system.
3. The (detached) car is subjected to three *external* forces. They are caused by:

   i. **It's weight**   ii. **Force in the cable**

   iii. **Reaction from surface**

4. Draw the free-body diagram of the (detached) car showing all these forces labeled with their magnitudes and directions. Include any other information e.g. angles, lengths etc. which may help when formulating the equations of motion.
5. Draw the corresponding kinetic diagram.

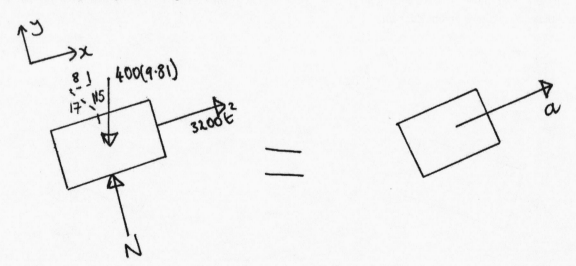

6. Using the $xy-axes$ system on the free-body diagram, write down the equation of motion in the x-direction:

$$\nearrow + \sum F_x = ma_x:\ 3200t^2 - 400(9.81)\left(\frac{8}{17}\right) = 400a$$

7. Solve for the acceleration of the car:

$$a(t) = (8t^2 - 4.616)\ \text{m/s}^2;\quad a(2) = 27.384\ \text{m/s}^2 \rightarrow \qquad \textbf{Ans.}$$

Problem 3.8

Block *A* has a weight of 8 lb and block *B* has a weight of 6 lb. They rest on a surface for which the coefficient of kinetic friction is $\mu_k = 0.2$. If the spring has a stiffness of $k = 20$ lb/ft, and it is compressed 0.2 ft, draw free-body diagrams for both blocks and use them to determine the acceleration of each block just after they are released.

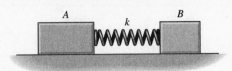

Solution

1. The size/shape of the blocks does not affect the motion under consideration. Consequently, we assume that the blocks have *negligible size* so that they can be modelled as particles.

2. Imagine each block to be separated or detached from the system.

3. Block *A* is subjected to four *external* forces. They are caused by:

   i.   **ii.**

   iii.   **iv.**

   Block *B* is subjected to four external forces. They are caused by:

   i.   **ii.**

   iii.   **iv.**

4. Draw the free-body diagrams of each (detached) block showing all these forces labeled with their magnitudes and directions. Include any other information e.g. angles, lengths etc. which may help when formulating the equations of motion. What is the direction of the acceleration vector for each block. Show this on a kinetic diagram or on the inertial coordinate system chosen in each free-body diagram.

A

B

5. Using the xy-axes system on the free-body diagram, write down the equation of motion in the x-direction for each block:

   Block *A*: $\overset{+}{\leftarrow} \sum F_x = ma_x$:

   Block *B*: $\overset{+}{\rightarrow} \sum F_x = ma_x$:

6. Solve for the acceleration in each case:

Problem 3.8

Block A has a weight of 8 lb and block B has a weight of 6 lb. They rest on a surface for which the coefficient of kinetic friction is $\mu_k = 0.2$. If the spring has a stiffness of $k = 20$ lb/ft, and it is compressed 0.2 ft, draw free-body diagrams for both blocks and use them to determine the acceleration of each block just after they are released.

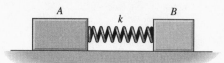

Solution

1. The size/shape of the blocks does not affect the motion under consideration. Consequently, we assume that the blocks have *negligible size* so that they can be modelled as particles.

2. Imagine each block to be separated or detached from the system.

3. Block A is subjected to four *external* forces. They are caused by:

   i. It's weight **ii. Spring force**

   iii. Friction **iv. Reaction from surface**

   Block B is subjected to four external forces. They are caused by:

   i. It's weight **ii. Spring force**

   iii. Friction **iv. Reaction from surface**

4. Draw the free-body diagrams of each (detached) block showing all these forces labeled with their magnitudes and directions. Include any other information e.g. angles, lengths etc. which may help when formulating the equations of motion. What is the direction of the acceleration vector for each block. Show this on a kinetic diagram or on the inertial coordinate system chosen in each free-body diagram.

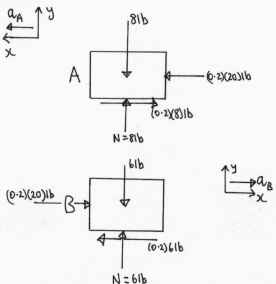

5. Using the xy-axes system on the free-body diagram, write down the equation of motion in the x-direction for each block:

   Block A: $\xleftarrow{+} \sum F_x = ma_x$: $4 - 1.6 = \dfrac{8}{32.2}a_A$

   Block B: $\xrightarrow{+} \sum F_x = ma_x$: $4 - 1.2 = \dfrac{6}{32.2}a_B$

6. Solve for the acceleration in each case:

$$a_A = 9.66 \text{ ft/s}^2 \leftarrow, \quad a_B = 15.0 \text{ ft/s}^2 \rightarrow$$ **Ans.**

Problem 3.9

When crossing an intersection, a motorcyclist encounters the slight bump or crown caused by the intersecting road. The crest of the bump has a radius of curvature of $\rho = 50$ ft. Draw free-body and kinetic diagrams for the motorcycle with rider. Use these diagrams to formulate equations of motion for the motorcycle with rider and find the maximum constant speed he can travel without leaving the surface of the road. Neglect the size of the motorcycle and the rider in the calculation. The rider and his motorcycle have a total weight of 450 lb.

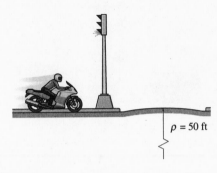

$\rho = 50$ ft

Solution

1. The motorcycle and rider have *negligible size* so that together they can be modelled as a particle.
2. Imagine the motorcycle and rider to be separated or detached from the system.
3. The (detached) motorcycle and rider is subjected to four *external* forces. They are caused by:

   i.   ii.

   iii.   iv.

4. Draw the free-body diagram of the (detached) motorcycle and rider (at the instant he encounters the bump) showing all these forces labeled with their magnitudes and directions. Include any other information e.g. angles, lengths etc. which may help when formulating the equations of motion. Which information given in the question suggests you use a $nt - coordinate$ system as the chosen inertial system? Show the corresponding acceleration components on a kinetic diagram or on the inertial coordinate system chosen in the free-body diagram.

5. Using the nt-axes system on the free-body diagram, write down the equation of motion in the n-direction:

   $$+ \downarrow \sum F_n = ma_n:$$

6. Solve for the acceleration component a_n under the appropriate conditions and hence the required speed v of the motorcycle:

Problem 3.9

When crossing an intersection, a motorcyclist encounters the slight bump or crown caused by the intersecting road. The crest of the bump has a radius of curvature of $\rho = 50$ ft. Draw free-body and kinetic diagrams for the motorcycle with rider. Use these diagrams to formulate equations of motion for the motorcycle with rider and find the maximum constant speed he can travel without leaving the surface of the road. Neglect the size of the motorcycle and the rider in the calculation. The rider and his motorcycle have a total weight of 450 lb.

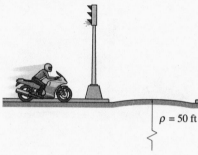

Solution

1. The motorcycle and rider have *negligible size* so that together they can be modelled as a particle.
2. Imagine the motorcycle and rider to be separated or detached from the system.
3. The (detached) motorcycle and rider is subjected to four *external* forces. They are caused by:

   i. Total weight   **ii. Friction at surface**

   iii. Drag   **iv. Reaction from surface**

4. Draw the free-body diagram of the (detached) motorcycle and rider (at the instant he encounters the bump) showing all these forces labeled with their magnitudes and directions. Include any other information e.g. angles, lengths etc. which may help when formulating the equations of motion. Which information given in the question suggests you use a $nt-coordinate$ system as the chosen inertial system? Show the corresponding acceleration components on a kinetic diagram or on the inertial coordinate system chosen in the free-body diagram.

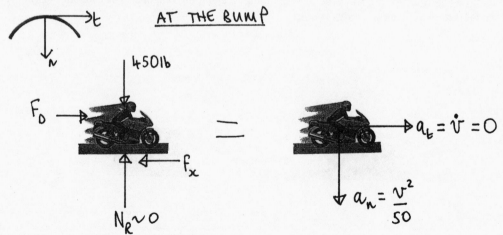

5. Using the nt-axes system on the free-body diagram, write down the equation of motion in the n-direction:

$$+\downarrow \sum F_n = ma_n: \quad 450 - N_R = \frac{450}{32.2}a_n$$

6. Solve for the acceleration component a_n under the appropriate conditions and hence the required speed v of the motorcycle:

$$\text{Let } N_R = 0 \text{ and } a_n = \frac{v^2}{50} \Rightarrow v = 40.1 \text{ ft/s} \qquad \textbf{Ans.}$$

Problem 3.10

The 2-kg spool S fits loosely on the inclined rod for which the coefficient of static friction is $\mu_s = 0.2$. If the spool is located 0.25 m from A, use a free-body diagram of the spool to determine the minimum constant speed the spool can have so that it does not slip down the rod.

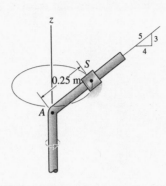

Solution

1. The spool has *negligible size* so that it can be modelled as a particle.
2. Imagine the spool to be separated or detached from the system.
3. The (detached) spool is subjected to three *external* forces. They are caused by:

   i.   ii.

   iii.

4. Draw the free-body diagram of the (detached) spool showing all these forces labeled with their magnitudes and directions. Include any other information e.g. angles, lengths etc. which may help when formulating the equations of motion. Which information given in the question suggests you use a *nt*-coordinate system as the chosen inertial system? Show the corresponding acceleration components on a kinetic diagram or on the inertial coordinate system chosen in the free-body diagram.

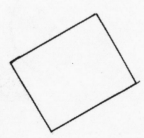

5. Using the *nt*-axes system on the free-body diagram, write down the equations of motion in the n and t-directions:

$$\xleftarrow{+} \sum F_n = ma_n:$$

$$+\uparrow \sum F_t = ma_t:$$

6. Solve for the required speed v of the spool:

Problem 3.10

The 2-kg spool S fits loosely on the inclined rod for which the coefficient of static friction is $\mu_s = 0.2$. If the spool is located 0.25 m from A, use a free-body diagram of the spool to determine the minimum constant speed the spool can have so that it does not slip down the rod.

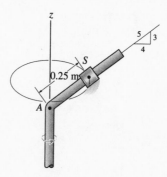

Solution

1. The spool has *negligible size* so that it can be modelled as a particle.
2. Imagine the spool to be separated or detached from the system.
3. The (detached) spool is subjected to three *external* forces. They are caused by:

      i. It's weight   **ii. Reaction from surface**

      iii. Friction

4. Draw the free-body diagram of the (detached) spool showing all these forces labeled with their magnitudes and directions. Include any other information e.g. angles, lengths etc. which may help when formulating the equations of motion. Which information given in the question suggests you use a $nt - coordinate$ system as the chosen inertial system? Show the corresponding acceleration components on a kinetic diagram or on the inertial coordinate system chosen in the free-body diagram.

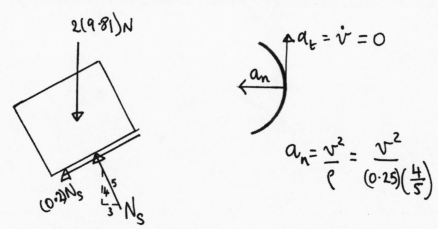

5. Using the nt-axes system on the free-body diagram, write down the equations of motion in the n and t-directions:

$$\xleftarrow{+} \sum F_n = ma_n: \quad N_s\left(\frac{3}{5}\right) + 0.2N_s\left(\frac{4}{5}\right) = 2a_n$$

$$+\uparrow \sum F_t = ma_t: \quad N_s\left(\frac{4}{5}\right) - 0.2N_s\left(\frac{3}{5}\right) - 2(9.81) = 2a_t$$

6. Solve for the required speed v of the spool:

Set $a_n = \dfrac{v^2}{0.2}$, $a_t = 0 (\dot v = 0)$ and obtain: $N_s = 28.85N$, $v = 1.48$ m/s   **Ans.**

3.2 Free-Body Diagrams in Rigid Body Kinetics

Problem 3.11

Draw the free-body and kinetic diagrams of the 2-lb bottle, with center of gravity at G, resting on the check-out conveyor. Use these diagrams to write down the equations of motion for the bottle.

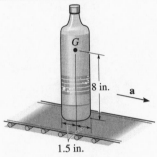

8 in.

a

1.5 in.

Solution

1. Imagine the bottle to be separated or detached from the system.
2. The bottle is subjected to three *external* forces (don't forget the weight!). They are caused by:

   i.   **ii.**

   iii.

3. Draw the free-body diagram of the (detached) bottle showing all these forces labeled with their magnitudes and directions. The line of action of the force of the belt on the bottle will vary depending on whether the bottle will slip or tip. This should be clear from your free-body diagram. Include any other relevant information e.g. lengths, angles etc. which may help when formulating the equations of motion (including the moment equation) for the bottle.
4. Draw the corresponding kinetic diagram.

5. Using the inertial coordinate system chosen on the free-body diagram and a suitably chosen point O, write down the equations of motion:

$$\xrightarrow{} + \sum F_x = m(a_G)_x :$$

$$+\uparrow \sum F_y = m(a_G)_y :$$

$$\circlearrowleft \rightarrow + \sum M_O = \sum (M_k)_O :$$

Problem 3.11

Draw the free-body and kinetic diagrams of the 2-lb bottle, with center of gravity at G, resting on the check-out conveyor. Use these diagrams to write down the equations of motion for the bottle.

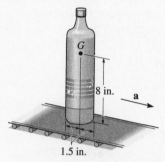

Solution

1. Imagine the bottle to be separated or detached from the system.
2. The bottle is subjected to three *external* forces (don't forget the weight!). They are caused by:

   i. It's weight   **ii. Friction**

   iii. Reaction from surface

3. Draw the free-body diagram of the (detached) bottle showing all these forces labeled with their magnitudes and directions. The line of action of the force of the belt on the bottle will vary depending on whether the bottle will slip or tip. This should be clear from your free-body diagram. Include any other relevant information e.g. lengths, angles etc. which may help when formulating the equations of motion (including the moment equation) for the bottle.
4. Draw the corresponding kinetic diagram.

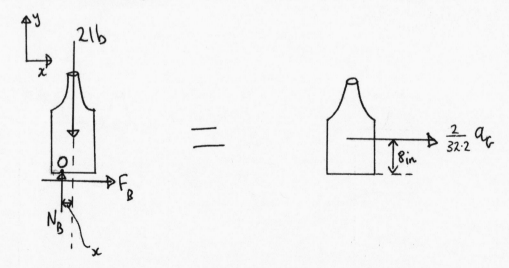

5. Using the inertial coordinate system chosen on the free-body diagram and a suitably chosen point O:

$$\rightarrow +\sum F_x = m(a_G)_x: \quad F_B = \frac{2}{32.2}a_G$$

$$+\uparrow \sum F_y = m(a_G)_y: \quad N_B 0 - 2 = 0$$

$$\circlearrowleft +\sum M_O = \sum(M_k)_O: \quad 2x = \frac{2}{32.2}a_G(8)$$

Problem 3.12

Draw the free-body and kinetic diagrams of the 200 lb door with center of gravity at G, if a man pushes on it at C with a horizontal force with magnitude F. There are rollers at A and B.

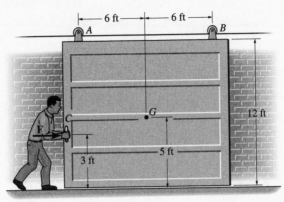

Solution

1. Imagine the door to be separated or detached from the system.
2. The door is subjected to four *external* forces (don't forget the weight!). They are caused by:

   i.   **ii.**

   iii.   **iv.**

3. Draw the free-body diagram of the (detached) door showing all these forces labeled with their magnitudes and directions. Include any other relevant information e.g. lengths, angles etc. which may help when formulating the equations of motion (including the moment equation) for the door.
4. Draw the corresponding kinetic diagram.

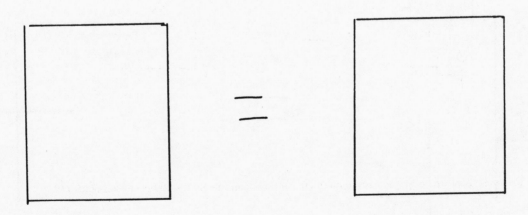

Problem 3.12

Draw the free-body and kinetic diagrams of the 200 lb door with center of gravity at G, if a man pushes on it at C with a horizontal force with magnitude F. There are rollers at A and B.

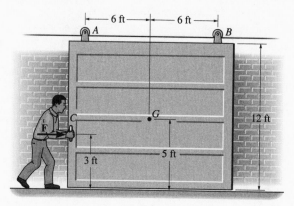

Solution

1. Imagine the door to be separated or detached from the system.
2. The door is subjected to four *external* forces (don't forget the weight!). They are caused by:

   i. It's weight

   ii. Roller at A

   iii. Roller at B

   iv. Force F

3. Draw the free-body diagram of the (detached) door showing all these forces labeled with their magnitudes and directions. Include any other relevant information e.g. lengths, angles etc. which may help when formulating the equations of motion (including the moment equation) for the door.
4. Draw the corresponding kinetic diagram.

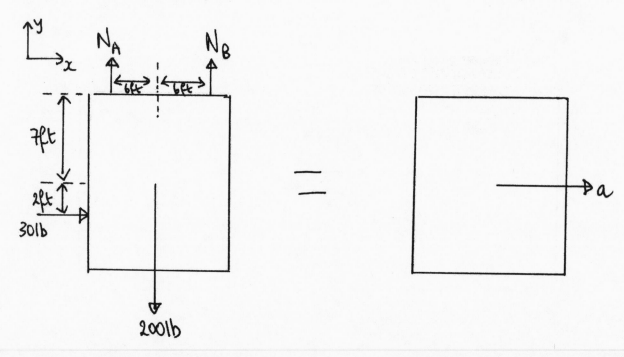

Problem 3.13

The jet has a total mass of 22 Mg and a center of mass at G. Initially at take-off, the engines provide a thrust $2T = 4$ kN and $T' = 1.5$ kN. Draw the free-body and kinetic diagrams of the jet. Neglect the mass of the wheels and, due to low velocity, neglect any lift caused by the wings. There are *two* wing wheels at B and one nose wheel at A.

Solution

1. Imagine the jet to be separated or detached from the system.
2. The jet is subjected to five *external* forces (don't forget the weight!). They are caused by:

   i.   ii.

   iii.   iv.

   v.

3. Draw the free-body diagram of the (detached) jet showing all these forces labeled with their magnitudes and directions. Include any other relevant information e.g. lengths, angles etc. which may help when formulating the equations of motion (including the moment equation) for the jet.
4. Draw the corresponding kinetic diagram.

Problem 3.13

The jet has a total mass of 22 Mg and a center of mass at G. Initially at take-off, the engines provide a thrust $2T = 4$ kN and $T' = 1.5$ kN. Draw the free-body and kinetic diagrams of the jet. Neglect the mass of the wheels and, due to low velocity, neglect any lift caused by the wings. There are *two* wing wheels at B and one nose wheel at A.

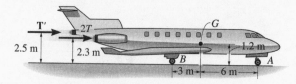

Solution

1. Imagine the jet to be separated or detached from the system.

2. The jet is subjected to five *external* forces (don't forget the weight!). They are caused by:

   i. It's weight   **ii. Reaction at A**

   iii. Reactions at B   **iv. Thrust of magnitude $2T$**

   v. Thrust T'

3. Draw the free-body diagram of the (detached) jet showing all these forces labeled with their magnitudes and directions. Include any other relevant information e.g. lengths, angles etc. which may help when formulating the equations of motion (including the moment equation) for the jet.

4. Draw the corresponding kinetic diagram.

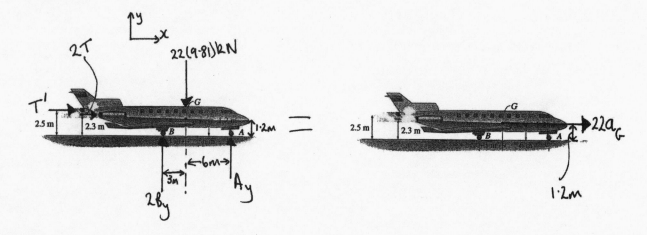

Problem 3.14

The top truck has a mass of 1.75 Mg and a center of mass at G. It is tied to the transport using a chain DE. The transport accelerates at 2 m/s. Draw the free-body and kinetic diagrams of the top truck.

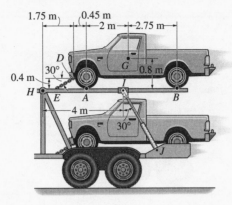

Solution

1. Imagine the top truck to be separated or detached from the system.
2. The truck is subjected to four *external* forces (don't forget the weight!). They are caused by:

   i.   ii.

   iii.   iv.

3. Draw the free-body diagram of the (detached) truck showing all these forces labeled with their magnitudes and directions. Include any other relevant information e.g. lengths, angles etc. which may help when formulating the equations of motion (including the moment equation) for the truck.
4. Draw the corresponding kinetic diagram.

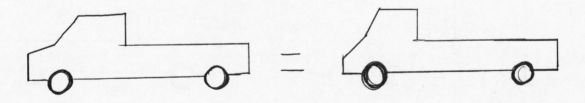

Problem 3.14

The top truck has a mass of 1.75 Mg and a center of mass at G. It is tied to the transport using a chain DE. The transport accelerates at 2 m/s. Draw the free-body and kinetic diagrams of the top truck.

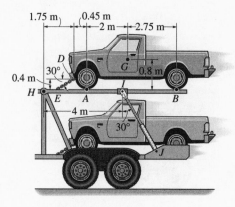

Solution

1. Imagine the top truck to be separated or detached from the system.

2. The truck is subjected to four *external* forces (don't forget the weight!). They are caused by:

    i. It's weight   **ii. Reactions at A**

    iii. Reactions at B   **iv. Chain DE**

3. Draw the free-body diagram of the (detached) truck showing all these forces labeled with their magnitudes and directions. Include any other relevant information e.g. lengths, angles etc. which may help when formulating the equations of motion (including the moment equation) for the truck.

4. Draw the corresponding kinetic diagram.

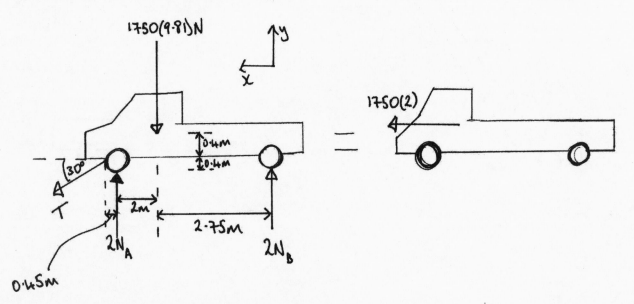

Problem 3.15

The drop gate at the end of the trailer has a mass of 1.25 Mg and mass center at G. It is supported by the cable AB and hinge at C. The truck begins to accelerate at 5 m/s^2. Draw the free-body and kinetic diagrams of the drop gate.

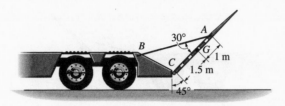

Solution

1. Imagine the drop gate to be separated or detached from the system.
2. The gate is subjected to four *external* forces (don't forget the weight!). They are caused by:

    i.   **ii.**

    iii.   **iv.**

3. Draw the free-body diagram of the (detached) gate showing all these forces labeled with their magnitudes and directions. Include any other relevant information e.g. lengths, angles etc. which may help when formulating the equations of motion (including the moment equation) for the gate.
4. Draw the corresponding kinetic diagram.

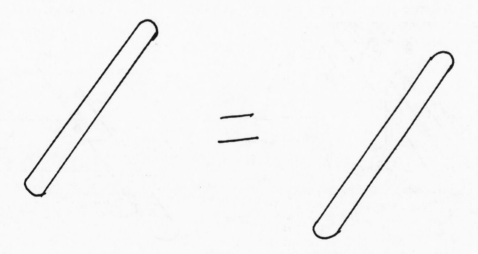

Problem 3.15

The drop gate at the end of the trailer has a mass of 1.25 Mg and mass center at G. It is supported by the cable AB and hinge at C. The truck begins to accelerate at 5 m/s^2. Draw the free-body and kinetic diagrams of the drop gate.

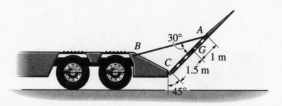

Solution

1. Imagine the drop gate to be separated or detached from the system.
2. The gate is subjected to four *external* forces (don't forget the weight!). They are caused by:

   i. It's weight   **ii.** <u>Two</u> reactions at C

   iii. Cable AB

3. Draw the free-body diagram of the (detached) gate showing all these forces labeled with their magnitudes and directions. Include any other relevant information e.g. lengths, angles etc. which may help when formulating the equations of motion (including the moment equation) for the gate.
4. Draw the corresponding kinetic diagram.

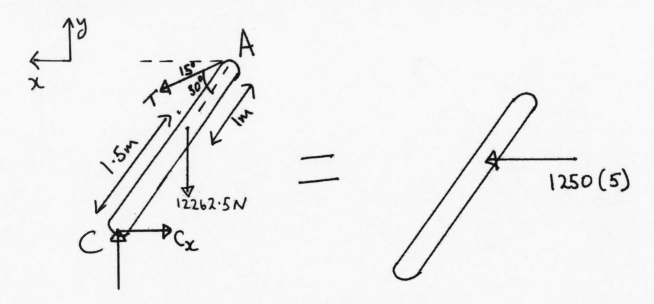

Problem 3.16

The sports car has a weight of 4500 lb and center of gravity at G. It starts fom rest causing the rear wheels to slip as it accelerates. The coefficient of kinetic friction at the road is $\mu_k = 0.3$. Draw the free-body and kinetic diagrams of the car. Neglect the mass of the wheels.

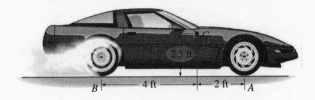

Solution

1. Imagine the car to be separated or detached from the system.

2. The car is subjected to four *external* forces. They are caused by:

       i.   **ii.**

       iii.   **iv.**

3. Draw the free-body diagram of the (detached) car showing all these forces labeled with their magnitudes and directions. Include any other relevant information e.g. lengths, angles etc. which may help when formulating the equations of motion (including the moment equation) for the car.

4. Draw the corresponding kinetic diagram.

Problem 3.16

The sports car has a weight of 4500 lb and center of gravity at G. It starts fom rest causing the rear wheels to slip as it accelerates. The coefficient of kinetic friction at the road is $\mu_k = 0.3$. Draw the free-body and kinetic diagrams of the car. Neglect the mass of the wheels.

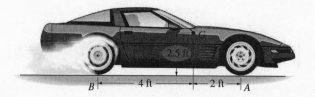

Solution

1. Imagine the car to be separated or detached from the system.
2. The car is subjected to four *external* forces. They are caused by:

    i. It's weight
    ii. Reactions at A
    iii. Reactions at B
    iv. Friction at rear wheels

3. Draw the free-body diagram of the (detached) car showing all these forces labeled with their magnitudes and directions. Include any other relevant information e.g. lengths, angles etc. which may help when formulating the equations of motion (including the moment equation) for the car.
4. Draw the corresponding kinetic diagram.

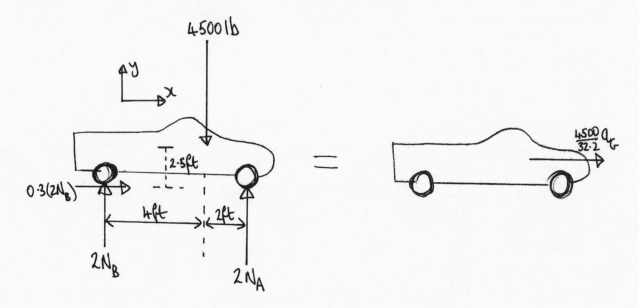

Problem 3.17

The drum truck supports the 600 lb drum that has a center of gravity at G. The operator pushes it forward with a horizontal force of 20 lb. Draw free-body and kinetic diagrams for the drum truck. Neglect the mass of the (4) wheels.

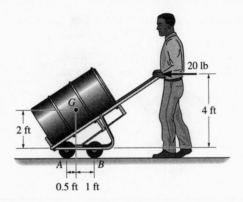

Solution

1. Imagine the drum truck to be separated or detached from the system.
2. The drum truck is subjected to four *external* forces. They are caused by:

   i.   ii.

   iii.   iv.

3. Draw the free-body diagram of the (detached) truck showing all these forces labeled with their magnitudes and directions. Include any other relevant information e.g. lengths, angles etc. which may help when formulating the equations of motion (including the moment equation) for the truck.
4. Draw the corresponding kinetic diagram.

Problem 3.17

The drum truck supports the 600 lb drum that has a center of gravity at G. The operator pushes it forward with a horizontal force of 20 lb. Draw free-body and kinetic diagrams for the drum truck. Neglect the mass of the (4) wheels.

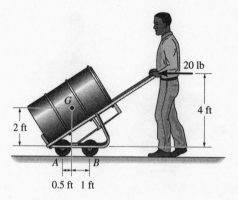

Solution

1. Imagine the drum truck to be separated or detached from the system.
2. The drum truck is subjected to four *external* forces. They are caused by:

    i. **It's weight**

    ii. **Reactions at A**

    iii. **Reactions at B**

    iv. **Force of magnitude 20 lb**

3. Draw the free-body diagram of the (detached) truckshowing all these forces labeled with their magnitudes and directions. Include any other relevant information e.g. lengths, angles etc. which may help when formulating the equations of motion (including the moment equation) for the truck.
4. Draw the corresponding kinetic diagram.

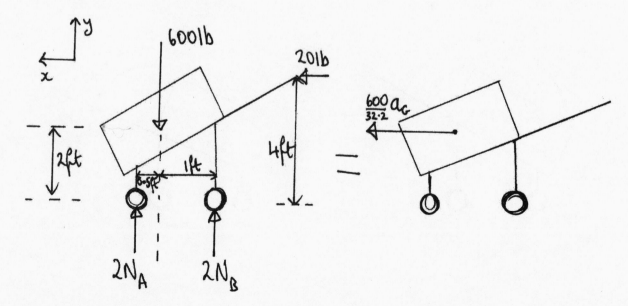

Problem 3.18

The arched pipe has a mass of 80 kg and rests on the surface of the platform. As it is hoisted from one level to the next, $\alpha = 0.25$ rad/s^2 and $\omega = 0.5$ rad/s at the instant $\theta = 30°$. The pipe does not slip. Draw the free-body and kinetic diagrams of the pipe at this instant.

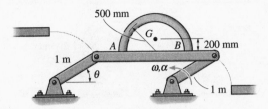

Solution

1. Imagine the pipe to be separated or detached from the system.

2. The pipe is subjected to five *external* forces. They are caused by:

    i.   ii.

    iii.   iv.

    v.

3. Draw the free-body diagram of the (detached) pipe showing all these forces labeled with their magnitudes and directions. Include any other relevant information e.g. lengths, angles etc. which may help when formulating the equations of motion (including the moment equation) for the pipe.

4. Draw the corresponding kinetic diagram.

Problem 3.18

The arched pipe has a mass of 80 kg and rests on the surface of the platform. As it is hoisted from one level to the next, $\alpha = 0.25$ rad/s^2 and $\omega = 0.5$ rad/s at the instant $\theta = 30°$. The pipe does not slip. Draw the free-body and kinetic diagrams of the pipe at this instant.

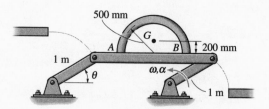

Solution

1. Imagine the pipe to be separated or detached from the system.
2. The pipe is subjected to five *external* forces. They are caused by:

   i. **It's weight**
   ii. **Reaction at A**
   iii. **Reaction at B**
   iv. **Friction at A**
   v. **Friction at B**

3. Draw the free-body diagram of the (detached) pipe showing all these forces labeled with their magnitudes and directions. Include any other relevant information e.g. lengths, angles etc. which may help when formulating the equations of motion (including the moment equation) for the pipe.
4. Draw the corresponding kinetic diagram.

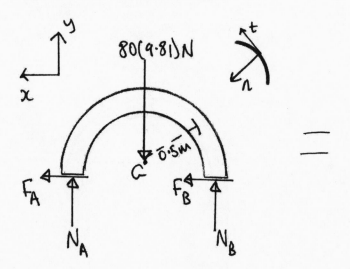

$$a_t = 80\,(0.25)(1) = 20$$

$$a_n = 80(0.5)^2\,(1) = 20$$

Problem 3.19

The desk has a weight of 75 lb and a center of gravity at G. A man pushes on it at C with a force with magnitude $F = 60$ lb. The coefficient of kinetic friction at A and B is $\mu_k = 0.2$. Draw the free-body and kinetic diagrams of the desk.

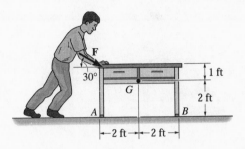

Solution

1. Imagine the desk to be separated or detached from the system.
2. The desk is subjected to six *external* forces. They are caused by:

   i.   ii.

   iii.   iv.

   v.   vi.

3. Draw the free-body diagram of the (detached) desk showing all these forces labeled with their magnitudes and directions. Include any other relevant information e.g. lengths, angles etc. which may help when formulating the equations of motion (including the moment equation) for the desk.
4. Draw the corresponding kinetic diagram.

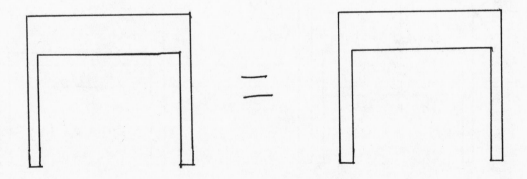

Problem 3.19

The desk has a weight of 75 lb and a center of gravity at G. A man pushes on it at C with a force with magnitude $F = 60$ lb. The coefficient of kinetic friction at A and B is $\mu_k = 0.2$. Draw the free-body and kinetic diagrams of the desk.

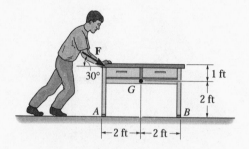

Solution

1. Imagine the desk to be separated or detached from the system.
2. The desk is subjected to six *external* forces. They are caused by:

       i. It's weight
       iii. Reaction at B
       v. Friction at B

       ii. Reaction at A
       iv. Friction at A
       vi. Applied force F

3. Draw the free-body diagram of the (detached) desk showing all these forces labeled with their magnitudes and directions. Include any other relevant information e.g. lengths, angles etc. which may help when formulating the equations of motion (including the moment equation) for the desk.
4. Draw the corresponding kinetic diagram.

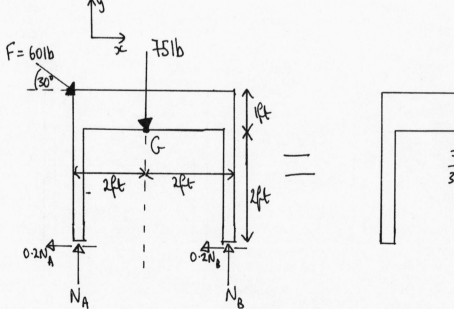

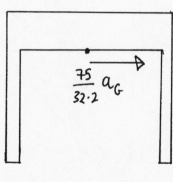

Problem 3.20

The smooth 180-lb pipe has a length of 20 ft and a negligible diameter. It is carried on a truck as shown. Draw the free-body and kinetic diagrams of the pipe.

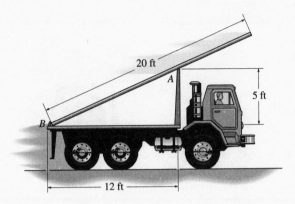

Solution

1. Imagine the pipe to be separated or detached from the system.
2. The pipe is subjected to four *external* forces. They are caused by:

   i.   **ii.**

   iii.   **iv.**

3. Draw the free-body diagram of the (detached) pipe showing all these forces labeled with their magnitudes and directions. Include any other relevant information e.g. lengths, angles etc. which may help when formulating the equations of motion (including the moment equation) for the pipe.
4. Draw the corresponding kinetic diagram.

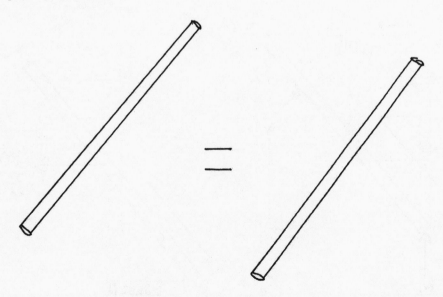

Problem 3.20

The smooth 180-lb pipe has a length of 20 ft and a negligible diameter. It is carried on a truck as shown. Draw the free-body and kinetic diagrams of the pipe.

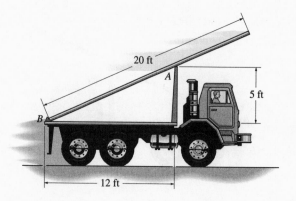

Solution

1. Imagine the pipe to be separated or detached from the system.
2. The pipe is subjected to four *external* forces. They are caused by:

   i. It's weight   **ii. Reaction at** *A*

   iii. Two reactions at *B*

3. Draw the free-body diagram of the (detached) pipe showing all these forces labeled with their magnitudes and directions. Include any other relevant information e.g. lengths, angles etc. which may help when formulating the equations of motion (including the moment equation) for the pipe.
4. Draw the corresponding kinetic diagram.

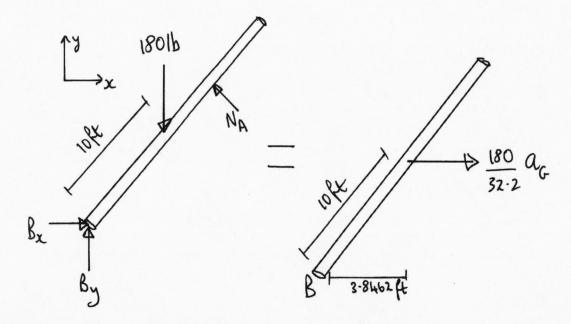

Problem 3.21

The van has a weight of 4500 lb and a center of gravity at G_v. It carries a fixed 800 lb load which has center of gravity at G_l. The van is travelling at 40 ft/s when the brakes are applied causing all the wheels to lock or skid. The coefficient of kinetic friction betwen the wheels and the pavement is $\mu_k = 0.3$. Draw the free-body and kinetic diagrams of the van. Neglect the mass of the wheels. Use these diagrams to write down equations of motion for the van and hence calculate the deceleration of the van as it skids to a complete stop.

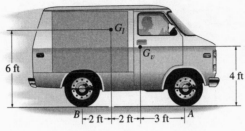

Solution

1. Imagine the van to be separated or detached from the system.

2. The van (including load) is subjected to six *external* forces. They are caused by:

   i.   **ii.**

   iii.   **iv.**

   v.   **vi.**

3. Draw the free-body diagram of the (detached) van showing all these forces labeled with their magnitudes and directions. Include any other relevant information e.g. lengths, angles etc. which may help when formulating the equations of motion (including the moment equation) for the van.

4. Draw the corresponding kinetic diagram.

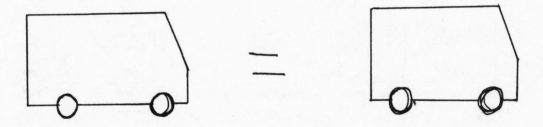

5. Using the inertial coordinate system chosen on the free-body diagram write down the equations of motion:

$$\leftarrow + \sum F_x = m(a_G)_x:$$
$$+ \uparrow \sum F_y = m(a_G)_y:$$

6. Solve for the acceleration of the van:

Problem 3.21.

 The van has a weight of 4500 lb and a center of gravity at G_v It carries a fixed 800 lb load which has center of gravity at G_l. The van is travelling at 40 ft/s when the brakes are applied causing all the wheels to lock or skid. The coefficient of kinetic friction betwen the wheels and the pavement is $\mu_k = 0.3$. Draw the free-body and kinetic diagrams of the van. Neglect the mass of the wheels. Use these diagrams to write down equations of motion for the van and hence calculate the deceleration of the van as it skids to a complete stop.

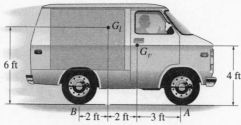

Solution

1. Imagine the van to be separated or detached from the system.
2. The van (including load) is subjected to six *external* forces. They are caused by:

   i. **It's weight**   ii. **Reaction at** A

   iii. **Reaction at** B iv. **Friction at** A

   v. **Friction at** B vi. **Weight of load** W_L

3. Draw the free-body diagram of the (detached) van showing all these forces labeled with their magnitudes and directions. Include any other relevant information e.g. lengths, angles etc. which may help when formulating the equations of motion (including the moment equation) for the van.
4. Draw the corresponding kinetic diagram.

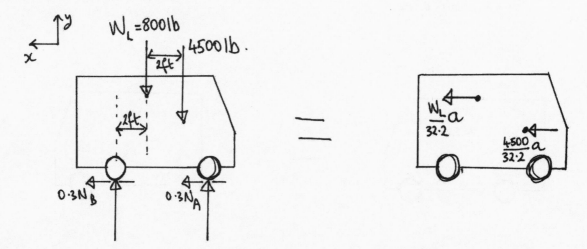

5. Using the inertial coordinate system chosen on the free-body diagram write down the equations of motion:

$$\leftarrow + \sum F_x = m(a_G)_x: \ 0.3N_B + 0.3N_A = \frac{W_L}{32.2}a + \frac{4500}{32.2}a$$

$$+ \uparrow \sum F_y = m(a_G)_y: \ N_B + N_A - W_L - 4500 = 0$$

6. Solve for the acceleration of the van:

$$\text{Set } W_L = 800 \text{ lb and obtain } \boldsymbol{a} = 9.66 \text{ ft/s}^2 \leftarrow$$

Ans.

Problem 3.22

The forks on the tractor support the pallet that carries a mass of 400 kg. The load, which is initially at rest, is subjected to curvilinear translation with a radius of 3 m as it is lowered with the maximum initial angular acceleration permitted to prevent it from slipping. The coefficient of static friction between the pallet and the forks is $\mu_s = 0.4$. Draw free-body and kinetic diagrams for the load. Use these diagrams to write down equations of motion for the load and hence calculate the required initial maximum angular acceleration.

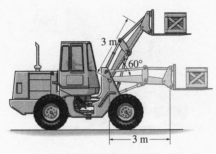

Solution

1. Imagine the load to be separated or detached from the system.
2. The load is subjected to three *external* forces. They are caused by:

    i.   **ii.**

    iii.

3. Draw the free-body diagram of the (detached) load showing all these forces labeled with their magnitudes and directions. Include any other relevant information e.g. lengths, angles etc. which may help when formulating the equations of motion (including the moment equation) for the load.
4. Draw the corresponding kinetic diagram.

5. Using the inertial coordinate system chosen on the free-body diagram write down the equations of motion:

    $$\rightarrow + \sum F_x = m(a_G)_x:$$
    $$+ \downarrow \sum F_y = m(a_G)_y:$$

6. Solve for the angular acceleration of the load:

Problem 3.22

The forks on the tractor support the pallet that carries a mass of 400 kg. The load, which is initially at rest, is subjected to curvilinear translation with a radius of 3 m as it is lowered with the maximum initial angular acceleration permitted to prevent it from slipping. The coefficient of static friction between the pallet and the forks is $\mu_s = 0.4$. Draw free-body and kinetic diagrams for the load. Use these diagrams to write down equations of motion for the load and hence calculate the required initial maximum angular acceleration.

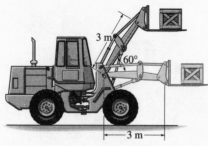

Solution

1. Imagine the load to be separated or detached from the system.
2. The load is subjected to three *external* forces. They are caused by:

    i. It's weight   **ii. Reaction from surface**

    iii. Friction at surface

3. Draw the free-body diagram of the (detached) load showing all these forces labeled with their magnitudes and directions. Include any other relevant information e.g. lengths, angles etc. which may help when formulating the equations of motion (including the moment equation) for the load.
4. Draw the corresponding kinetic diagram.

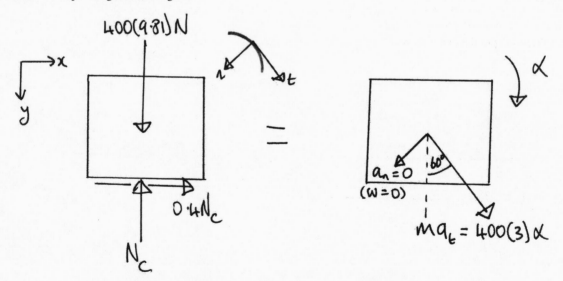

5. Using the inertial coordinate system chosen on the free-body diagram write down the equations of motion:

$$\rightarrow + \sum F_x = m(a_G)_x: \quad 0.4N_C = 400(3)\alpha \sin 60°$$

$$+ \downarrow \sum F_y = m(a_G)_y: \quad -N_C + 400(9.81) = 400(3)\cos 60°$$

6. Solve for the angular acceleration of the load:

$$N_C = 3.19kN, \quad \alpha = 1.23 \text{ rad/s}^2 \curvearrowright$$

                                                                               Ans.

Problem 3.23

The uniform bar BC has a weight of 40 lb and is pin-connected to the two links which have negligible mass. Draw free-body and kinetic diagrams for the bar BC at the instant $\theta = 30°$.

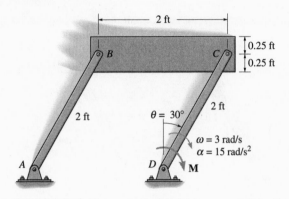

Solution

1. Imagine the bar to be separated or detached from the system.
2. The bar is subjected to five *external* forces. They are caused by:

   i.   ii.

   iii.   iv.

   v.

3. Draw the free-body diagram of the (detached) bar showing all these forces labeled with their magnitudes and directions. Include any other relevant information e.g. lengths, angles etc. which may help when formulating the equations of motion (including the moment equation) for the pipe.
4. Draw the corresponding kinetic diagram.

Problem 3.23

The uniform bar BC has a weight of 40 lb and is pin-connected to the two links which have negligible mass. Draw free-body and kinetic diagrams for the bar BC at the instant $\theta = 30°$.

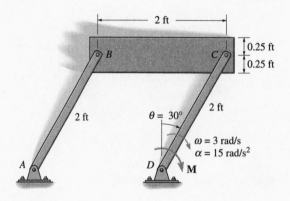

Solution

1. Imagine the bar to be separated or detached from the system.
2. The bar is subjected to five *external* forces. They are caused by:

      i. It's weight   **ii. Two reactions at** C

      iii. Two reactions at B

3. Draw the free-body diagram of the (detached) bar showing all these forces labeled with their magnitudes and directions. Include any other relevant information e.g. lengths, angles etc. which may help when formulating the equations of motion (including the moment equation) for the pipe.
4. Draw the corresponding kinetic diagram.

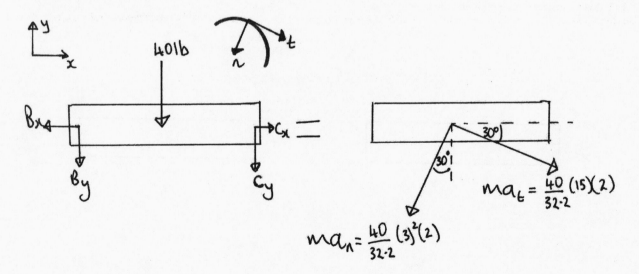

Problem 3.24

The 80 kg disk is supported by a pin at A and is rotating clockwise at $\omega = 0.5$ rad/s when it is in the position shown. Draw the free-body and kinetic diagrams of the disk at this instant.

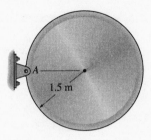

Solution

1. Imagine the disk to be separated or detached from the system.
2. The disk is subjected to three *external* forces. They are caused by:

   i.   ii.

   iii.

3. Draw the free-body diagram of the (detached) disk showing all these forces labeled with their magnitudes and directions. Include any other relevant information e.g. lengths, angles etc. which may help when formulating the equations of motion (including the moment equation) for the disk.
4. Draw the corresponding kinetic diagram indicating clearly the acceleration components of the disk.

Problem 3.24

The 80 kg disk is supported by a pin at A and is rotating clockwise at $\omega = 0.5$ rad/s when it is in the position shown. Draw the free-body and kinetic diagrams of the disk at this instant.

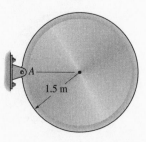

Solution

1. Imagine the disk to be separated or detached from the system.
2. The disk is subjected to three *external* forces. They are caused by:

   i. It's weight   **ii. Two reactions at A**

3. Draw the free-body diagram of the (detached) disk showing all these forces labeled with their magnitudes and directions. Include any other relevant information e.g. lengths, angles etc. which may help when formulating the equations of motion (including the moment equation) for the disk.
4. Draw the corresponding kinetic diagram indicating clearly the acceleration components of the disk.

$$80(9\cdot81)N$$

$$A_x$$

$$A_y$$

$$G \qquad 1\cdot5m$$

$$I_G \alpha$$

$$A \xleftarrow{a_n} G \quad \downarrow a_t$$

$$=$$

$$ma_n = (80)\omega^2(1\cdot5) = (80)(0\cdot5)^2(1\cdot5)$$

$$ma_t = 80(\alpha)(1\cdot5)$$

Problem 3.25

The drum has a weight of 20 lb and a radius of gyration about its mass center of 0.8 ft. If the block has a weight of 12 lb, draw the free-body and kinetic diagrams of the drum and block system and use them to determine the angular acceleration of the drum if the block is allowed to fall freely.

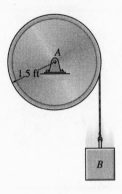

Solution

1. Imagine the drum and block system to be separated or detached from the pin at A.

2. The drum and block system is subjected to four *external* forces. They are caused by:

   i.   **ii.**

   iii.   **iv.**

3. Draw the free-body diagram of the (detached) drum and block system showing all these forces labeled with their magnitudes and directions. Include any other relevant information e.g. lengths, angles etc. which may help when formulating the equations of motion (including the moment equation) for the drum and block system.

4. Draw the corresponding kinetic diagram indicating clearly the acceleration components of the drum and block.

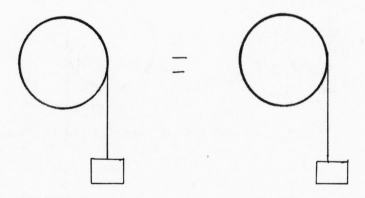

5. Sum moments about A (why?) and write down the moment equation of motion:

$$\circlearrowleft + \sum M_A = \sum (M_k)_A:$$

6. Solve for the angular acceleration α_D of the drum:

Problem 3.25

The drum has a weight of 20 lb and a radius of gyration about its mass center of 0.8 ft. If the block has a weight of 12 lb, draw the free-body and kinetic diagrams of the drum and block system and use them to determine the angular acceleration of the drum if the block is allowed to fall freely.

Solution

1. Imagine the drum and block system to be separated or detached from the pin at A.
2. The drum and block system is subjected to four *external* forces. They are caused by:

   i. Weight of drum   **ii. Two reactions at A**

   iii. Weight of block

3. Draw the free-body diagram of the (detached) drum and block system showing all these forces labeled with their magnitudes and directions. Include any other relevant information e.g. lengths, angles etc. which may help when formulating the equations of motion (including the moment equation) for the drum and block system.
4. Draw the corresponding kinetic diagram indicating clearly the acceleration components of the drum and block.

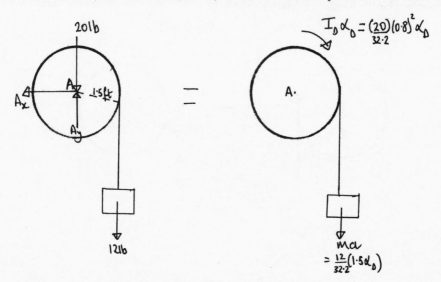

5. Sum moments about A (why? — **eliminates pin reactions**) and write down the moment equation of motion:

$$\circlearrowleft \; + \sum M_A = \sum (M_k)_A: \quad 12(1.5) = \left[\frac{20}{32.2}(0.8)^2 \right]\alpha_D + \left[\frac{12}{32.2}(1.5\alpha_D) \right](1.5)$$

6. Solve for the angular acceleration α_D of the drum:

$$\alpha_D = 14.6 \text{ rad/s}^2 \; \curvearrowleft \qquad\qquad\qquad\qquad \textbf{Ans.}$$

Problem 3.26

The 10 lb bar is pinned at its center O and connected to a torsional spring. The spring has a stiffness $k = 5$ lb.ft/rad so that the torque developed is $M = (5\theta)$ lb.ft, where θ is in radians. The bar is released from rest when $\theta = 90°$. Draw a free-body diagram for the bar when $\theta = 45°$.

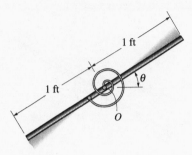

Solution

1. Imagine the bar to be separated or detached from the system.
2. The bar is subjected to three *external* forces and one applied couple moment. They are caused by:

   i.   **ii.**

   iii.   **iv.**

3. Draw the free-body diagram of the (detached) bar showing all these forces labeled with their magnitudes and directions. Include any other relevant information e.g. lengths, angles etc. which may help when formulating the equations of motion (including the moment equation) for the bar.
4. Indicate the acceleration components of the bar on the coordinate axes system chosen in the free-body diagram.

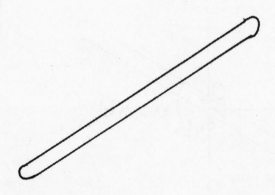

Problem 3.26

The 10 lb bar is pinned at its center O and connected to a torsional spring. The spring has a stiffness $k = 5$ lb.ft/rad so that the torque developed is $M = (5\theta)$ lb.ft, where θ is in radians. The bar is released from rest when $\theta = 90°$. Draw a free-body diagram for the bar when $\theta = 45°$.

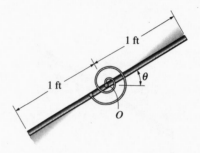

Solution

1. Imagine the bar to be separated or detached from the system.
2. The bar is subjected to three *external* forces and one applied couple moment. They are caused by:

    i. It's weight   **ii. <u>Two</u> reactions at O**

    iii. Applied torque M

3. Draw the free-body diagram of the (detached) bar showing all these forces labeled with their magnitudes and directions. Include any other relevant information e.g. lengths, angles etc. which may help when formulating the equations of motion (including the moment equation) for the bar.
4. Indicate the acceleration components of the bar on the coordinate axes system chosen in the free-body diagram.

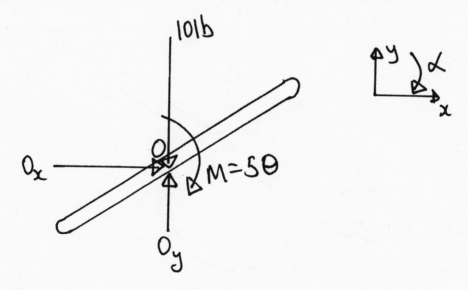

Problem 3.27

The 20 kg roll of paper has a radius of gyration $k_A = 90$ mm about an axis passing through point A. It is pin-supported at both ends by two brackets AB. The roll rests against a wall for which the coefficient of kinetic friction is $\mu_k = 0.2$. A constant vertical force of magnitude F is applied to the roll to pull off 1 m of paper starting from rest. Draw a free-body diagram for the roll of paper. Neglect the mass of paper that is removed.

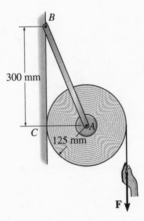

Solution

1. Imagine the roll to be separated or detached from the system.
2. The bar is subjected to five *external* forces (use one force to descibe the effect of AB on the roll) They are caused by:

   i.   ii.

   iii.   iv.

   v.

3. Draw the free-body diagram of the (detached) roll showing all these forces labeled with their magnitudes and directions. Include any other relevant information e.g. lengths, angles etc. which may help when formulating the equations of motion (including the moment equation) for the roll.

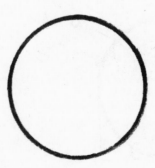

Problem 3.27

The 20 kg roll of paper has a radius of gyration $k_A = 90$ mm about an axis passing through point A. It is pin-supported at both ends by two brackets AB. The roll rests against a wall for which the coefficient of kinetic friction is $\mu_k = 0.2$. A constant vertical force of magnitude F is applied to the roll to pull off 1 m of paper starting from rest. Draw a free-body diagram for the roll of paper. Neglect the mass of paper that is removed.

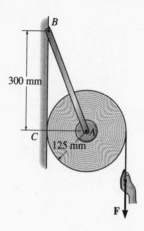

Solution

1. Imagine the roll to be separated or detached from the system.
2. The bar is subjected to five *external* forces (use one force to descibe the effect of AB on the roll) They are caused by:

   i. **It's weight**

   ii. **Reaction at C**

   iii. **Friction at C**

   iv. **Force in bar AB**

   v. **Applied force F**

3. Draw the free-body diagram of the (detached) roll showing all these forces labeled with their magnitudes and directions. Include any other relevant information e.g. lengths, angles etc. which may help when formulating the equations of motion (including the moment equation) for the roll.

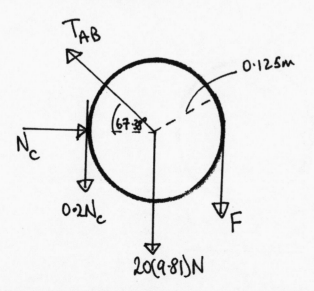

Problem 3.28

The door will close automatically using torsional springs mounted on the hinges. Each spring has a stiffness $k = 50$ N.m/rad so that the torque on each hinge is $M = (50\theta)$ N.m where θ is measured in radians. The door is released from rest when it is open at $\theta = 90°$. Treating the door as a thin plate having a mass of 70 kg, draw a free-body diagram for the door at the instant $\theta = 0°$.

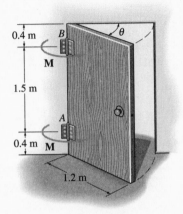

Solution

1. Imagine the door to be separated or detached from the system.
2. The door is subjected to five *external* forces and an applied couple moment. They are caused by:

   i.   ii.

   iii.   iv.

   v.   vi.

3. Draw the free-body diagram of the (detached) door showing all these forces labeled with their magnitudes and directions. Include any other relevant information e.g. lengths, angles etc. which may help when formulating the equations of motion (including the moment equation) for the door.

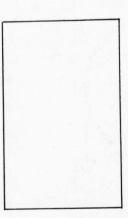

Problem 3.28

The door will close automatically using torsional springs mounted on the hinges. Each spring has a stiffness $k = 50$ N.m/rad so that the torque on each hinge is $M = (50\theta)$ N.m where θ is measured in radians. The door is released from rest when it is open at $\theta = 90°$. Treating the door as a thin plate having a mass of 70 kg, draw a free-body diagram for the door at the instant $\theta = 0°$.

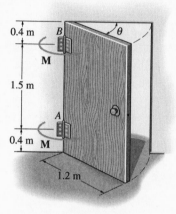

Solution

1. Imagine the door to be separated or detached from the system.
2. The door is subjected to five *external* forces and an applied couple moment. They are caused by:

   i. It's weight   **ii. Two reactions at A**

   iii. Two reactions at B   **iv. Applied torque**

3. Draw the free-body diagram of the (detached) door showing all these forces labeled with their magnitudes and directions. Include any other relevant information e.g. lengths, angles etc. which may help when formulating the equations of motion (including the moment equation) for the door.

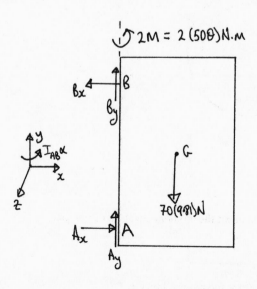

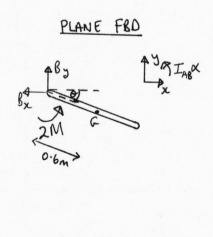

Problem 3.29

Cable is unwound from a spool supported on small rollers at A and B by exerting a force $T = 300$ N on the cable. The spool and cable have a total mass of 600 kg and a radius of gyration of $k_o = 1.2$ m. Draw a free-body diagram for the spool and use it to compute the time needed to unravel 5 m of cable from the spool. Neglect the mass of the cable being unwound and the mass of the rollers at A and B. The rollers turn with no friction.

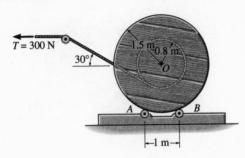

Solution

1. Imagine the spool to be separated or detached from the system.
2. The spool is subjected to four *external* forces. They are caused by:

   i.   **ii.**

   iii.   **iv.**

3. Draw the free-body diagram of the (detached) spool showing all these forces labeled with their magnitudes and directions. Include any other relevant information e.g. lengths, angles etc. which may help when formulating the equations of motion (including the moment equation) for the spool.

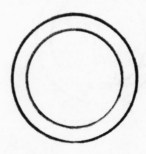

4. Sum moments about the center of the spool (O) and write down the moment equation of motion:

$$\circlearrowleft + \sum M_O = I_O\alpha:$$

5. Solve for the angular acceleration α of the spool.
6. Use kinematics (in terms of angular displacement) to solve for the required time t.

Problem 3.29

Cable is unwound from a spool supported on small rollers at A and B by exerting a force $T = 300$ N on the cable. The spool and cable have a total mass of 600 kg and a radius of gyration of $k_o = 1.2$ m. Draw a free-body diagram for the spool and use it to compute the time needed to unravel 5 m of cable from the spool. Neglect the mass of the cable being unwound and the mass of the rollers at A and B. The rollers turn with no friction.

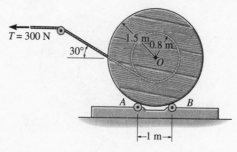

Solution

1. Imagine the spool to be separated or detached from the system.
2. The spool is subjected to four *external* forces. They are caused by:

   i. It's weight   **ii. Reaction at A**

   iii. Reaction at B   **iv. Applied force T**

3. Draw the free-body diagram of the (detached) spool showing all these forces labeled with their magnitudes and directions. Include any other relevant information e.g. lengths, angles etc. which may help when formulating the equations of motion (including the moment equation) for the spool.

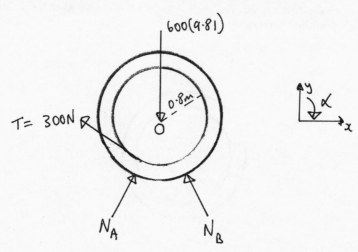

4. Sum moments about the center of the spool (O) and write down the moment equation of motion:

$$\curvearrowleft + \sum M_O = I_O\alpha: \quad 300(0.8) = 600(1.2)^2\alpha$$

5. Solve for the angular acceleration α of the spool: $\alpha = 0.2778\,rad/s^2 \curvearrowright$.
6. Use kinematics (in terms of angular displacement θ) to solve for the required time t.

$$\theta = \theta_0 + \omega_0 t + \frac{1}{2}\alpha t^2. \quad \text{Set } \theta = \frac{s}{r} = \frac{5}{0.8} = 6.25 \text{ rad.}$$

$$6.25 = 0 + 0 + \frac{1}{2}(0.2778)t^2 \Rightarrow t = 6.71 \text{ s}$$ **Ans.**

Problem 3.30

The disk has a mass M and a radius R. If a block of mass m is attached to the cord, draw free-body and kinetic diagrams of the disk and mass system and use them to determine the angular acceleration of the disk when the block is released from rest.

Solution

1. Imagine the disk and block system to be separated or detached from the pin at the center of the disk.

2. The disk and block system is subjected to four *external* forces. They are caused by:

   i.   **ii.**

   iii.   **iv.**

3. Draw the free-body diagram of the (detached) disk and block system showing all these forces labeled with their magnitudes and directions. Include any other relevant information e.g. lengths, angles etc. which may help when formulating the equations of motion (including the moment equation) for the disk and block system.

4. Draw the corresponding kinetic diagram indicating clearly the acceleration components of the disk and block.

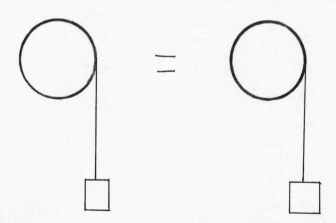

5. Sum moments about the center of the disk (O) (why?) and write down the moment equation of motion:

$$\circlearrowleft \;+\sum M_O = \sum (M_k)_O:$$

6. Solve for the angular acceleration α_D of the disk:

Problem 3.30

The disk has a mass M and a radius R. If a block of mass m is attached to the cord, draw free-body and kinetic diagrams of the disk and mass system and use them to determine the angular acceleration of the disk when the block is released from rest.

Solution

1. Imagine the disk and block system to be separated or detached from the pin at the center of the disk.
2. The disk and block system is subjected to four *external* forces. They are caused by:

   i. Weight of disk   **ii. Two reactions at A**

   iii. Weight of block

3. Draw the free-body diagram of the (detached) disk and block system showing all these forces labeled with their magnitudes and directions. Include any other relevant information e.g. lengths, angles etc. which may help when formulating the equations of motion (including the moment equation) for the disk and block system.
4. Draw the corresponding kinetic diagram indicating clearly the acceleration components of the disk and block.

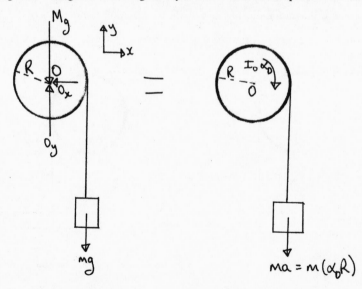

5. Sum moments about the center of the disk (O) (why? — **eliminates pin reactions**) and write down the moment equation of motion:

$$\circlearrowleft + \sum M_O = \sum (M_k)_O: \quad mgR = \frac{1}{2}MR^2(\alpha_D) + m(\alpha_D R)R$$

6. Solve for the angular acceleration α_D of the disk:

$$\alpha_D = \frac{2mg}{R(M + 2m)} \quad \circlearrowleft$$

Ans.

Problem 3.31

The 25 kg diving board is uniform and rigid. At the instant the man jumps off, the spring is compressed a maximum amount of 200 mm, $\omega = 0$ and the board is horizontal. Draw free-body and kinetic diagrams for the board at the instant the man jumps off. Take $k = 7$ kN/m.

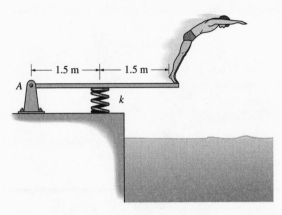

Solution

1. Imagine the board to be separated or detached from the system.

2. The board is subjected to four *external* forces. They are caused by:

   i.   **ii.**

   iii.   **iv.**

3. Draw the free-body diagram of the (detached) board showing all these forces labeled with their magnitudes and directions. Include any other relevant information e.g. lengths, angles etc. which may help when formulating the equations of motion (including the moment equation) for the board.

4. Draw the corresponding kinetic diagram. Be sure to indicate the components of the vectors $m\mathbf{a}$ and $I\boldsymbol{\alpha}$.

Problem 3.31

The 25 kg diving board is uniform and rigid. At the instant the man jumps off, the spring is compressed a maximum amount of 200 mm, $\omega = 0$ and the board is horizontal. Draw free-body and kinetic diagrams for the board at the instant the man jumps off. Take $k = 7$ kN/m.

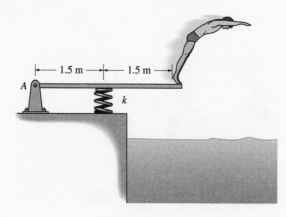

Solution

1. Imagine the board to be separated or detached from the system.
2. The board is subjected to four *external* forces. They are caused by:

   i. **Weight of board**   ii. **Two reactions at** A

   iii. **Spring force**

3. Draw the free-body diagram of the (detached) board showing all these forces labeled with their magnitudes and directions. Include any other relevant information e.g. lengths, angles etc. which may help when formulating the equations of motion (including the moment equation) for the board.

4. Draw the corresponding kinetic diagram. Be sure to indicate the components of the vectors $m\mathbf{a}$ and $I\alpha$.

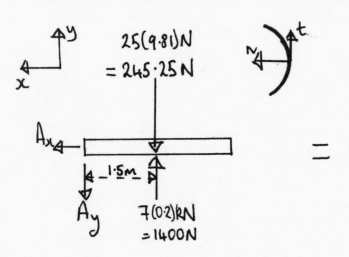

Problem 3.32

The two blocks A and B have mass m_A and m_B, respectively, where $m_B > m_A$. The pulley can be treated as a disk of mass M. Regarding the pulley and two blocks as a single system, draw the free-body and kinetic diagrams of this system and use them to formulate the appropriate equation of motion which, when solved, will lead to the acceleration of block A. Neglect the mass of the cord and any slipping on the pulley.

Solution

1. Imagine the system consisting of the pulley and blocks to be separated or detached.
2. This detached system is subjected to four *external* forces. They are caused by:

   i.   ii.

   iii.   iv.

3. Draw the free-body diagram of the (detached) system showing all these forces labeled with their magnitudes and directions. Include any other relevant information e.g. lengths, angles etc. which may help when formulating the equation of motion for this system.
4. Draw the corresponding kinetic diagram indicating clearly the components of the vectors $m\mathbf{a}$ and $I\boldsymbol{\alpha}$ for the (detached) system.

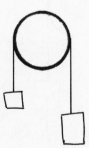 $=$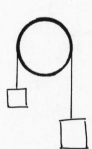

5. Sum moments about the center of the pulley (O) (why?) and write down the moment equation of motion:

$$\curvearrowleft + \sum M_O = \sum (M_k)_O:$$

6. Solve for the magnitude of the angular acceleration α of the pulley:
7. Find the required acceleration of block A:

Problem 3.32

The two blocks A and B have mass m_A and m_B, respectively, where $m_B > m_A$. The pulley can be treated as a disk of mass M. Regarding the pulley and two blocks as a single system, draw the free-body and kinetic diagrams of this system and use them to formulate the appropriate equation of motion which, when solved, will lead to the acceleration of block A. Neglect the mass of the cord and any slipping on the pulley.

Solution

1. Imagine the system consisting of the pulley and blocks to be separated or detached.

2. This detached system is subjected to four *external* forces. They are caused by:

   i. Weight of drum   **ii. Reactions at O**

   iii. Weight of blocks A and B

3. Draw the free-body diagram of the (detached) system showing all these forces labeled with their magnitudes and directions. Include any other relevant information e.g. lengths, angles etc. which may help when formulating the equation of motion for this system.

4. Draw the corresponding kinetic diagram indicating clearly the components of the vectors $m\mathbf{a}$ and $I\boldsymbol{\alpha}$ for the (detached) system.

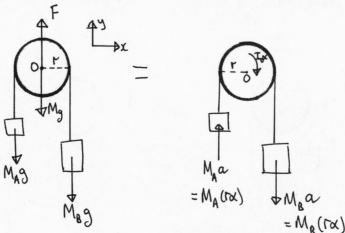

5. Sum moments about the center of the pulley (O) (why? — **eliminates two forces**) and write down the moment equation of motion: $\circlearrowleft + \sum M_O = \sum (M_k)_O$: $M_B g(r) - M_A g(r) = \left(\frac{1}{2} M r^2\right) \alpha + M_B r^2 \alpha + M_A r^2 \alpha$

6. Solve for the magnitude of the angular acceleration α of the pulley: $\alpha = \dfrac{g(M_B - M_A)}{r\left(\dfrac{1}{2}M + M_B + M_A\right)}$

7. Find the required acceleration of block A: $a = \dfrac{\alpha}{r} = \dfrac{g(M_B - M_A)}{\left(\dfrac{1}{2}M + M_B + M_A\right)}$, direction ↑.

Problem 3.33

The two-bar assembly is released from rest in the position shown. Each bar has a mass m and a length l. Joint B is a fixed joint. Draw the free-body and kinetic diagrams for the segment BC.

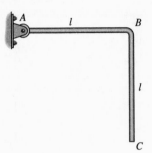

Solution

1. Imagine the member BC to be separated or detached from the system.
2. Determine the number and types of reactions *acting on the member* at B.
3. Draw the free-body diagram of the (detached) member showing all the external forces and moments *acting on the member* labeled with their magnitudes and directions. *Assume* the sense of the vectors representing the *reactions acting on the member*. Include any other relevant information e.g. lengths, angles etc. which may help when formulating the equations of motion for BC.
4. Draw the corresponding kinetic diagram for BC indicating clearly the components of the vectors $m\mathbf{a}$ and $I\alpha$ for the segment BC.

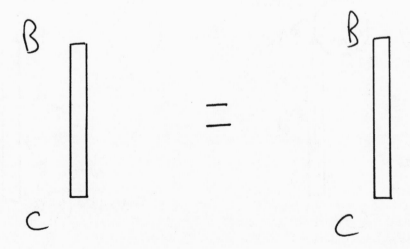

Problem 3.33

The two-bar assembly is released from rest in the position shown. Each bar has a mass m and a length l. Joint B is a fixed joint. Draw the free-body and kinetic diagrams for the segment BC.

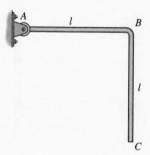

Solution

1. Imagine the member BC to be separated or detached from the system.
2. Determine the number and types of reactions *acting on the member* at B.
3. Draw the free-body diagram of the (detached) member showing all the external forces and moments *acting on the member* labeled with their magnitudes and directions. *Assume* the sense of the vectors representing the *reactions acting on the member*. Include any other relevant information e.g. lengths, angles etc. which may help when formulating the equations of motion for BC.
4. Draw the corresponding kinetic diagram for BC indicating clearly the components of the vectors $m\mathbf{a}$ and $I\boldsymbol{\alpha}$ for the segment BC.

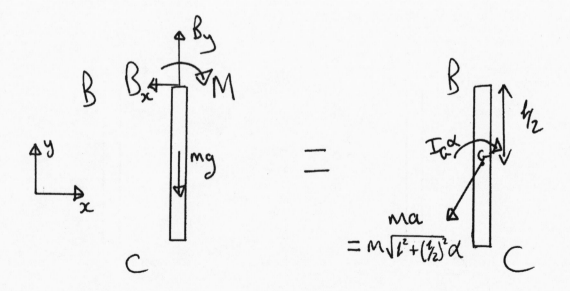

Problem 3.34

The wheel has a mass of 25 kg. It is originally spinning at $\omega = \omega_1$. It is placed on the ground, for which the coefficient of kinetic friction is $\mu_C = 0.5$, until the motion stops. Draw a free-body diagram for the wheel during this time.

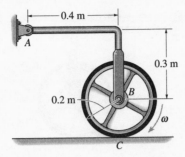

Solution

1. Imagine the wheel to be separated or detached from the system.
2. Determine the number and types of external forces *acting on the wheel*.
3. Draw the free-body diagram of the (detached) wheel showing all the external forces and moments *acting on the wheel* labeled with their magnitudes and directions. *Assume* the sense of the vectors representing the *reactions acting on the wheel*. Include any other relevant information e.g. lengths, angles etc. which may help when formulating the equations of motion for the wheel.
4. Indicate clearly the components of the vectors $m\mathbf{a}_G$ and $I_G\alpha$ for the wheel either on a separate kinetic diagram or on the coordinate system chosen for the free-body diagram.

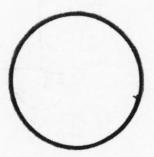

Problem 3.34

The wheel has a mass of 25 kg. It is originally spinning at $\omega = \omega_1$. It is placed on the ground, for which the coefficient of kinetic friction is $\mu_C = 0.5$, until the motion stops. Draw a free-body diagram for the wheel during this time.

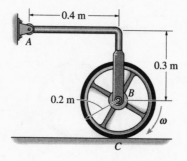

Solution

1. Imagine the wheel to be separated or detached from the system.
2. Determine the number and types of external forces *acting on the wheel*.
3. Draw the free-body diagram of the (detached) wheel showing all the external forces and moments *acting on the wheel* labeled with their magnitudes and directions. *Assume* the sense of the vectors representing the *reactions acting on the wheel*. Include any other relevant information e.g. lengths, angles etc. which may help when formulating the equations of motion for the wheel.
4. Indicate clearly the components of the vectors $m\mathbf{a}_G$ and $I_G\alpha$ for the wheel either on a separate kinetic diagram or on the coordinate system chosen for the free-body diagram.

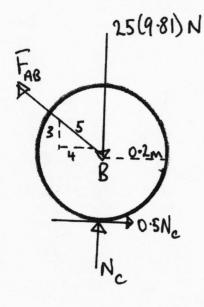

$$M(a_G)_x = M(a_G)_y = 0$$

Problem 3.35

A 40 kg boy sits on top of the large wheel which has a mass of 400 kg and a radius of gyration $k_G = 5.5$ m. The boy essentially starts from rest at $\theta = 0°$ and the wheel begins to rotate freely. Draw free-body and kinetic diagrams for the wheel regarding the boy as an attached particle. Draw also, a free-body diagram for the boy when he reaches the angle θ at which he begins to slide.

Solution

1. Imagine the wheel together with the boy (as a particle) to be separated or detached from the system.

2. There are four *external* forces acting: They are caused by:

    i.   **ii.**

    iii.   **iv.**

3. Draw the free-body diagram of the (detached) wheel showing all these forces labeled with their magnitudes and directions. Include any other relevant information e.g. lengths, angles etc. which may help when formulating the equations of motion for the wheel and the boy.

4. Draw the corresponding kinetic diagram. Be sure to indicate the components of the vectors $m\mathbf{a}$ and $I\alpha$.

5. Detach the boy from the entire system and draw his free-body diagram noting that there are three external forces acting on the boy.

 =

Problem 3.35

A 40 kg boy sits on top of the large wheel which has a mass of 400 kg and a radius of gyration $k_G = 5.5$ m. The boy essentially starts from rest at $\theta = 0°$ and the wheel begins to rotate freely. Draw free-body and kinetic diagrams for the wheel regarding the boy as an attached particle. Draw also, a free-body diagram for the boy when he reaches the angle θ at which he begins to slide.

Solution

1. Imagine the wheel together with the boy (as a particle) to be separated or detached from the system.

2. There are four *external* forces acting: They are caused by:

   i. Weight of wheel **ii. Weight of boy**

   iii. Two reactions at O

3. Draw the free-body diagram of the (detached) wheel showing all these forces labeled with their magnitudes and directions. Include any other relevant information e.g. lengths, angles etc. which may help when formulating the equations of motion for the wheel and the boy.

4. Draw the corresponding kinetic diagram. Be sure to indicate the components of the vectors $m\mathbf{a}$ and $I\boldsymbol{\alpha}$.

5. Detach the boy from the entire system and draw his free-body diagram noting that there are three external forces acting on the boy.

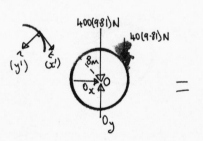

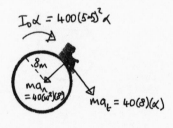

Problem 3.36

The slender rod of length L and mass m is released from rest when $\theta = 0°$. Using free-body and kinetic diagrams of the rod, formulate appropriate equations of motion and, determine, as a function of θ, the normal and frictional forces which are exerted on the ledge at A as it falls downward.

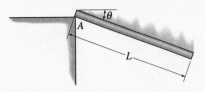

Solution

1. Imagine the rod to be separated or detached from the system.
2. The rod is subjected to three *external* forces. They are caused by:

   i.   ii.

   iii.

3. Draw the free-body diagram of the (detached) rod showing all these forces labeled with their magnitudes and directions. Include any other relevant information e.g. lengths, angles etc. which may help when formulating the equations of motion (including the moment equation) for the rod.
4. Draw the corresponding kinetic diagram indicating the components of the vectors $m\mathbf{a}_G$ and $I_G\alpha$.

5. Using the inertial coordinate system chosen on the free-body diagram write down the equations of motion:

   $$+\nwarrow \sum F_x = m(a_G)_x:$$
   $$+\swarrow \sum F_y = m(a_G)_y:$$
   $$\curvearrowright + \sum M_A = I_A\alpha:$$

6. Solve the third equation for α:
7. Integrate to obtain ω:
8. Use the first two equations of motion to find the required normal and frictional forces.

Problem 3.36

The slender rod of length L and mass m is released from rest when $\theta = 0°$. Using free-body and kinetic diagrams of the rod, formulate appropriate equations of motion and, determine, as a function of θ, the normal and frictional forces which are exerted on the ledge at A as it falls downward.

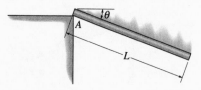

Solution

1. Imagine the rod to be separated or detached from the system.
2. The rod is subjected to three *external* forces. They are caused by:

    i. Weight of rod   **ii. Reaction at** A

    iii. Friction at A

3. Draw the free-body diagram of the (detached) rod showing all these forces labeled with their magnitudes and directions. Include any other relevant information e.g. lengths, angles etc. which may help when formulating the equations of motion (including the moment equation) for the rod.
4. Draw the corresponding kinetic diagram indicating the components of the vectors $m\mathbf{a}_G$ and $I_G\boldsymbol{\alpha}$.

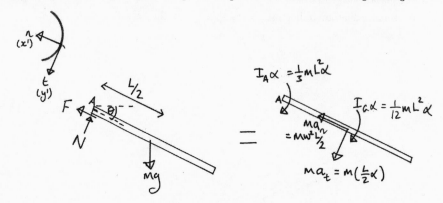

5. Using the inertial coordinate system chosen on the free-body diagram write down the equations of motion:

$$+\nwarrow \sum F_x = m(a_G)_x: \quad F - mg\sin\theta = m\omega^2\left(\frac{L}{2}\right)$$

$$+\swarrow \sum F_y = m(a_G)_y: \quad -N + mg\cos\theta = m\left(\frac{L}{2}\right)\alpha$$

$$\curvearrowright +\sum M_A = I_A\alpha: \quad mg\left(\frac{L}{2}\right)\cos\theta = \frac{1}{3}mL^2\alpha$$

6. Solve the third equation for α:

$$\alpha = 1.5\left(\frac{g}{L}\right)\cos\theta$$

7. Integrate to obtain ω:

$$\omega^2 = 3\left(\frac{g}{L}\right)\sin\theta$$

8. Use the first two equations of motion to find the required normal and frictional forces.

$$F = 2.5\,mg\sin\theta; \quad N = 0.25mg\cos\theta \quad \text{(directions as shown)}$$

Ans.

Problem 3.37

A force **F** of magnitude $F = 2$ lb is applied perpendicular to the axis of the 5 lb rod and moves from O to A at a constant rate of 4 ft/s. The rod is at rest when $\theta = 0$ and **F** is at O when $t = 0$. Draw a free-body diagram for the rod. Explain the significance of each force on the diagram.

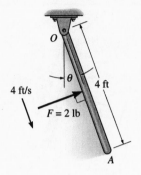

Solution

Problem 3.37

A force **F** of magnitude $F = 2$ lb is applied perpendicular to the axis of the 5lb rod and moves from O to A at a constant rate of 4 ft/s. The rod is at rest when $\theta = 0$ and **F** is at O when $t = 0$. Draw a free-body diagram for the rod. Explain the significance of each force on the diagram.

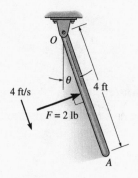

Solution

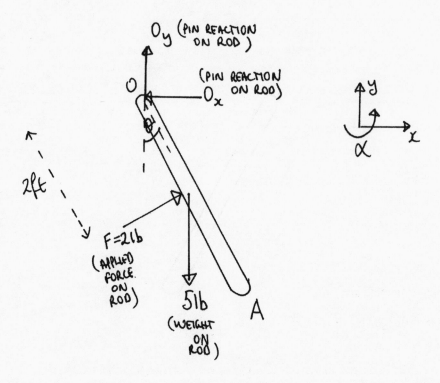

Problem 3.38

The support at B is suddenly removed. Draw free-body and kinetic diagrams for the rod ACB. Segments AC and CB each have a weight of 10 lb. Use the diagrams to formulate equations of motion for ACB. Hence find the initial horizontal and vertical components of reaction which the pin A exerts on the rod ACB.

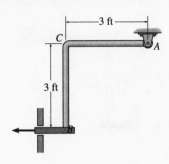

Solution

First draw the free-body and kinetic diagrams for ACB

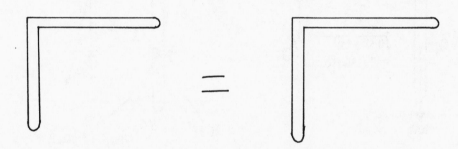

Next, write down the equations of motion:

1. Sum moments about A (why?):

$$\curvearrowleft + \sum M_A = I_A \alpha:$$

$$\left(I_A = \frac{1}{3}\left(\frac{10}{32.2}\right)(3)^2 + \frac{1}{12}\left(\frac{10}{32.2}\right)(3)^2 + \left(\frac{10}{32.2}\right)(1.5^2 + 3^2) = 4.6584 \text{ slug.ft}^2 \right)$$

2. Solve for α:

3. Write down the other two equations of motion:

$$\longrightarrow + \sum F_x = m(a_G)_x:$$

$$+ \downarrow \sum F_y = m(a_G)_y:$$

4. Using the value of α obtained above, solve for the required reaction components.

Problem 3.38

The support at B is suddenly removed. Draw free-body and kinetic diagrams for the rod ACB. Segments AC and CB each have a weight of 10 lb. Use the diagrams to formulate equations of motion for ACB. Hence find the initial horizontal and vertical components of reaction which the pin A exerts on the rod ACB.

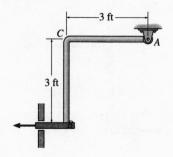

Solution

First draw the free-body and kinetic diagrams for ACB

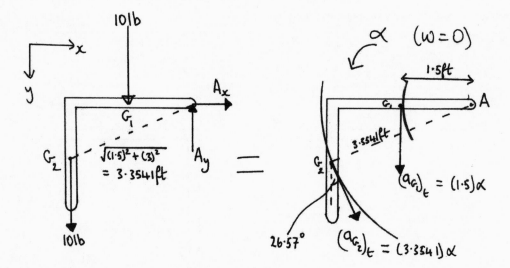

Next, write down the equations of motion:

1. Sum moments about A (why? — **Eliminate pin reactions**):

$$\curvearrowleft + \sum M_A = I_A\alpha: \quad 10(3) + 10(1.5) = 4.6584\alpha$$

$$\left(I_A = \frac{1}{3}\left(\frac{10}{32.2}\right)(3)^2 + \frac{1}{12}\left(\frac{10}{32.2}\right)(3)^2 + \left(\frac{10}{32.2}\right)(1.5^2 + 3^2) = 4.6584 \text{ slug.ft}^2\right)$$

2. Solve for α: $\alpha = 9.66$ rad/s^2
3. Write down the other two equations of motion:

$$\longrightarrow + \sum F_x = m(a_G)_x: \quad A_x = \left(\frac{10}{32.2}\right)\alpha(3.3541)\sin 26.57°$$

$$+\downarrow \sum F_y = m(a_G)_y: \quad 10 + 10 - A_y = \left(\frac{10}{32.2}\right)\alpha(1.5) + \left(\frac{10}{32.2}\right)\alpha(3.3541)\cos 26.57°$$

4. Using the value of α obtained above, solve for the required reaction components.

$$A_x = 4.50 \text{ lb}; \quad A_y = 6.50 \text{ lb} \qquad \textbf{Ans.}$$

Problem 3.39

The ladder is released from rest when $\theta = 0$. Treating the ladder as a slender rod of length l and mass m, draw free-body and kinetic diagrams for the ladder as it falls.

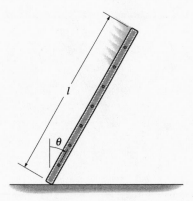

Solution

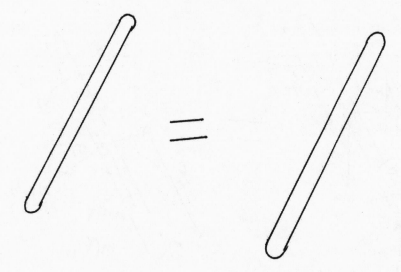

Problem 3.39

The ladder is released from rest when $\theta = 0$. Treating the ladder as a slender rod of length l and mass m, draw free-body and kinetic diagrams for the ladder as it falls.

Solution

Problem 3.40

The 20 kg punching bag is initially at rest and then subjected to a horizontal force of magnitude $F = 30$ N. Draw free-body and kinetic diagrams for the bag.

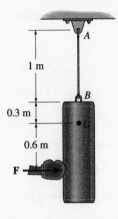

Solution

Problem 3.40

The 20 kg punching bag is initially at rest and then subjected to a horizontal force of magnitude $F = 30$ N. Draw free-body and kinetic diagrams for the bag.

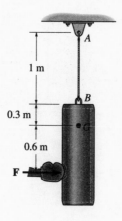

Solution

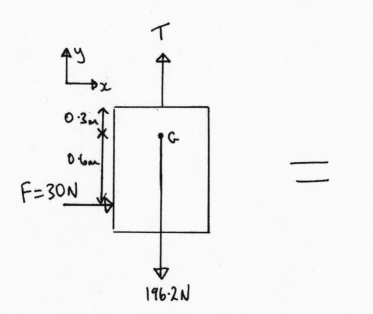

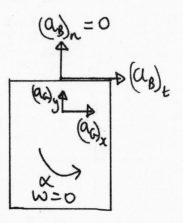

Problem 3.41

The spool has a mass of 500 kg. It rests on the surface of a conveyor belt for which the coefficient of kinetic friction is $\mu_k = 0.4$. The conveyor accelerates at 1 m/s^2 and the spool is originally at rest. Draw free-body and kinetic diagrams for the spool.

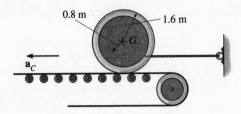

Solution

Problem 3.41

The spool has a mass of 500 kg. It rests on the surface of a conveyor belt for which the coefficient of kinetic friction is $\mu_k = 0.4$. The conveyor accelerates at 1 m/s^2 and the spool is originally at rest. Draw free-body and kinetic diagrams for the spool.

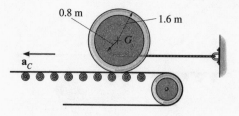

Solution

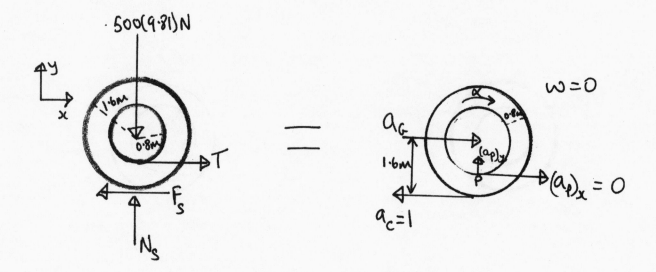

Problem 3.42

The upper body of the crash dummy has a mass of 75 lb, and a center of gravity at G. By means of the seat belt, this body segment is assumed to be pin-conected to the seat of the car at A. A crash causes the car to decelerate at 50 ft/s². Draw free-body and kinetic diagrams for the dummy when it has rotated $\theta°$.

Solution

Problem 3.42

The upper body of the crash dummy has a mass of 75 lb, and a center of gravity at G. By means of the seat belt, this body segment is assumed to be pin-conected to the seat of the car at A. A crash causes the car to decelerate at 50 ft/s². Draw free-body and kinetic diagrams for the dummy when it has rotated $\theta°$.

Solution

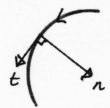

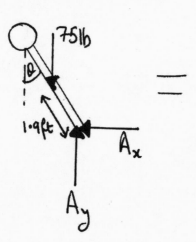

Problem 3.43

The 2 kg slender bar is supported by cord BC and then released from rest at A. Draw free-body and kinetic diagrams for the bar as it is released.

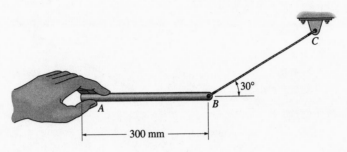

Solution

Problem 3.43

The 2 kg slender bar is supported by cord BC and then released from rest at A. Draw free-body and kinetic diagrams for the bar as it is released.

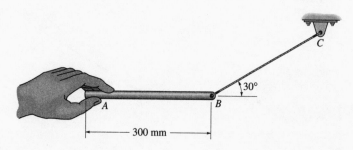

Solution

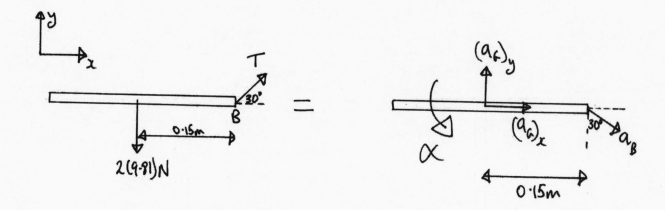

Problem 3.44

The uniform bar of mass m and length L is balanced in the vertical position when the horizontal force **P** is applied to the roller at A. Draw free-body and kinetic diagrams for the bar. Neglect the mass of the roller.

Solution

Problem 3.44

The uniform bar of mass m and length L is balanced in the vertical position when the horizontal force **P** is applied to the roller at A. Draw free-body and kinetic diagrams for the bar. Neglect the mass of the roller.

Solution

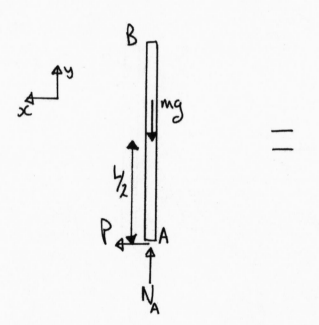

$=$

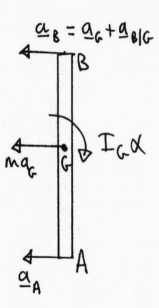

Problem 3.45

The ladder has a weight W and rests against the smooth wall and ground. It is released and allowed to slide downward. Treat the ladder as a slender rod and draw free-body and kinetic diagrams for the ladder.

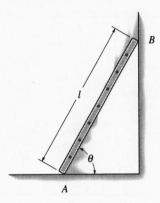

Solution

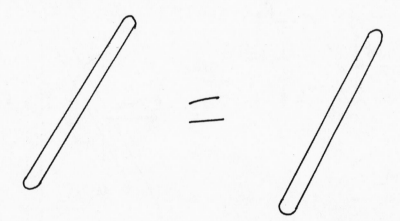

Problem 3.45

The ladder has a weight W and rests against the smooth wall and ground. It is released and allowed to slide downward. Treat the ladder as a slender rod and draw free-body and kinetic diagrams for the ladder.

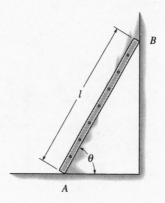

Solution

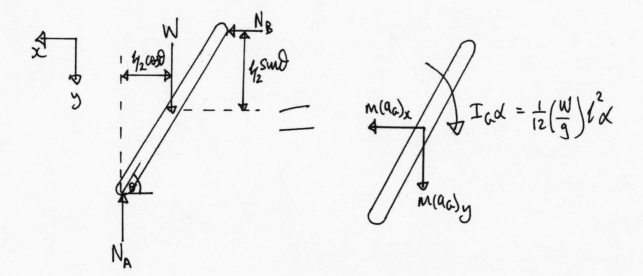

Problem 3.46

The 10-lb hoop or thin ring is given an initial angular velocity of 6 rad/s when it is placed on the surface. If the coefficient of kinetic friction between the hoop and the surface is $\mu_k = 0.3$, draw free-body and kinetic diagrams for the hoop as it slips. Use the diagrams to formulate equations of motion for the hoop. Hence find the normal force extered by the surface on the hoop, the angular acceleration of the hoop and the acceleration of its mass center.

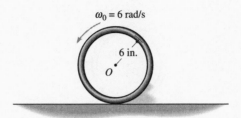

$\omega_0 = 6$ rad/s

6 in.

O

Solution

First draw the free-body and kinetic diagrams for the hoop.

Next, write down the equations of motion:

1. Write down the two translational equations of motion:

$$+\uparrow \sum F_y = m(a_G)_y:$$
$$\leftarrow + \sum F_x = m(a_G)_x:$$

2. Sum moments about O (why?):

$$\curvearrowright + \sum M_O = I_O\alpha:$$

3. Solve for the required quantities:

Problem 3.46

The 10-lb hoop or thin ring is given an initial angular velocity of 6 rad/s when it is placed on the surface. If the coefficient of kinetic friction between the hoop and the surface is $\mu_k = 0.3$, draw free-body and kinetic diagrams for the hoop as it slips. Use the diagrams to formulate equations of motion for the hoop. Hence find the normal force extered by the surface on the hoop, the angular acceleration of the hoop and the acceleration of its mass center.

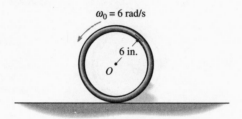

Solution

First draw the free-body and kinetic diagrams for the hoop.

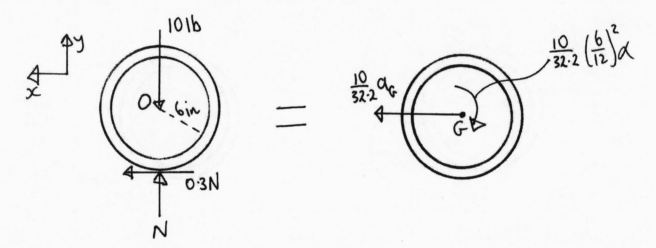

Next, write down the equations of motion:

1. Write down the two translational equations of motion:

$$+\uparrow \sum F_y = m(a_G)_y: \quad N - 10 = 0$$

$$\leftarrow +\sum F_x = m(a_G)_x: \quad 0.3N = \left(\frac{10}{32.2}\right)a_G$$

2. Sum moments about O (why? — **eliminates weight and normal reaction**):

$$\curvearrowright +\sum M_O = I_O\alpha: \quad (0.3)(N)\left(\frac{6}{12}\right) = \left(\frac{10}{32.2}\right)\left(\frac{6}{12}\right)^2 \alpha$$

3. Solve for the required quantities:

$$\mathbf{N} = 10 \text{ lb} \uparrow, \quad a_G = 9.66 \text{ ft/s}^2 \leftarrow, \quad \alpha = 19.32 \text{ rad/s}^2 \curvearrowright \qquad \text{Ans.}$$

Problem 3.47

A cord is wrapped around each of the two 10-kg disks. The disks are released from rest. Draw free-body and kinetic diagrams for each disk. Neglect the mass of the cord.

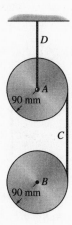

Solution

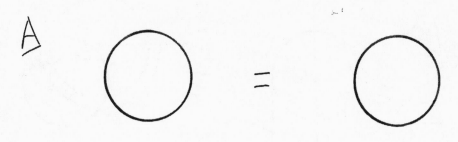

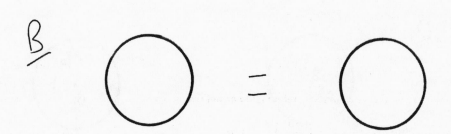

Problem 3.47

A cord is wrapped around each of the two 10-kg disks. The disks are released from rest. Draw free-body and kinetic diagrams for each disk. Neglect the mass of the cord.

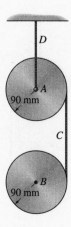

Solution

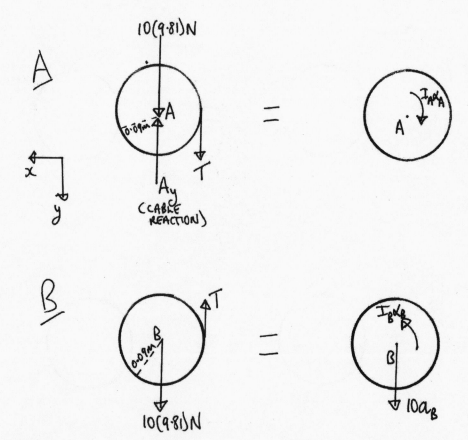

Problem 3.48

The 500-lb beam is supported at A and B when it is subjected to a force of 1000 lb as shown. The pin support at A suddenly fails. Draw free-body and kinetic diagrams for the beam. Assume the beam is a slender rod so that its thickness can be neglected.

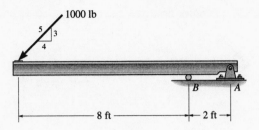

Solution

Problem 3.48

The 500-lb beam is supported at A and B when it is subjected to a force of 1000 lb as shown. The pin support at A suddenly fails. Draw free-body and kinetic diagrams for the beam. Assume the beam is a slender rod so that its thickness can be neglected.

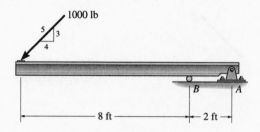

Solution

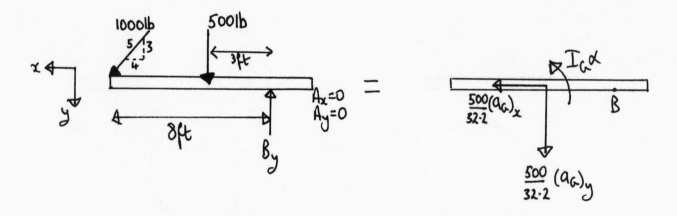

Problem 3.49

The 15-lb disk rests on the 5-lb plate. A cord is wrapped around the periphery of the disk and attached to the wall at B. A torque $M = 40$ lb.ft is applied to the disk. The disk does not slip on the plate and the surface at at D is smooth. Draw free-body and kinetic diagrams for the disk. Draw also a free-body diagram for the plate. Neglect the mass of the cord.

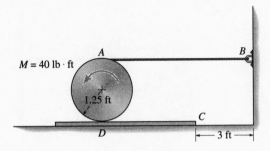

Solution

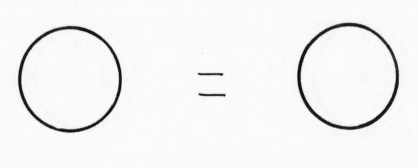

Problem 3.49

The 15-lb disk rests on the 5-lb plate. A cord is wrapped around the periphery of the disk and attached to the wall at B. A torque $M = 40$ lb.ft is applied to the disk. The disk does not slip on the plate and the surface at at D is smooth. Draw free-body and kinetic diagrams for the disk. Draw also a free-body diagram for the plate. Neglect the mass of the cord.

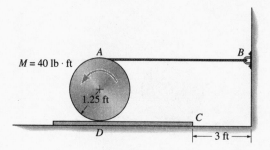

$M = 40$ lb · ft

A B

1.25 ft

C

D — 3 ft —

Solution

y

x

DISK

$15 lb$

T

$40 lb.ft$

G

$1.25 ft$

F_P

N_P

$=$

$I_G \alpha$

$\dfrac{15}{32.2} a_G$

PLATE

N_P F_P

SMOOTH

$5 lb$

N_f

Problem 3.50

A concrete pipe (thin ring) has a mass of 500 kg and radius of 0.5 m and rolls without slipping down a 300 kg ramp. The ramp is free to move. Draw free-body and kinetic diagrams for the ramp and the pipe.

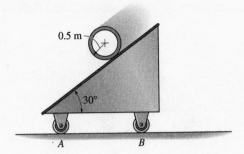

Solution

RAMP

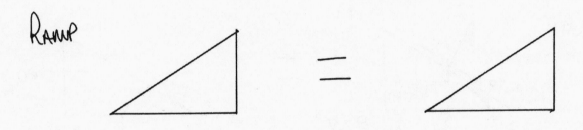

PIPE

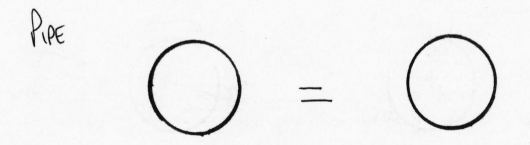

Problem 3.50

A concrete pipe (thin ring) has a mass of 500 kg and radius of 0.5 m and rolls without slipping down a 300 kg ramp. The ramp is free to move. Draw free-body and kinetic diagrams for the ramp and the pipe.

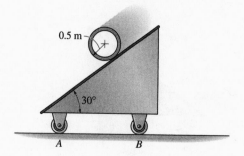

Solution

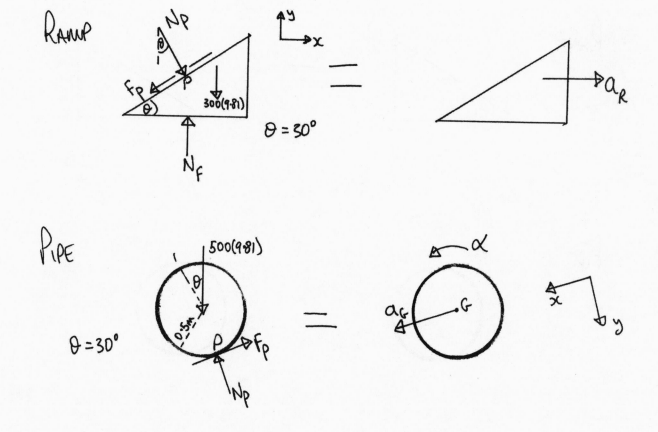